Basic

Astronomy

&

The Small

Telescope

Updated 2/2021

Matthew DeSipio Jr.

The Northern Hemisphere Night Sky

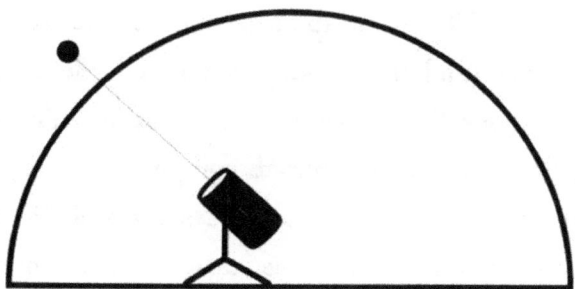

WARNING: Attempt to use and modify your own telescope at your own risk.

WARNING: Do NOT look at the sun with your telescope. You will damage your eyes (and the telescope's optics) unless you take the necessary precautions. Do not observe the sun until you have done your research and have all the appropriate equipment. "Safe" solar filters have a transmittance of 0.003% or lower. Colored and LPR/UHC filters transmit much more light than this, making them unsafe for solar observations. Eyepiece solar filters are dangerous — do not use them.

Table of Contents

Introduction

Astronomy is an incredible hobby for people of all ages, cultures, and backgrounds. This is because, for the most part, we are all fascinated by the night sky. You just bought a book about the night sky written by an author you know absolutely nothing about, so I am going to assume that you are particularly interested in it.

I will try to accomplish several things throughout the course of this book. The first and most important goal I have is to get you even more excited about astronomy (you have NO idea what you are getting yourself into) than you already are. I also want to convince you that you can absolutely, positively, and definitely use a small telescope to make astronomical observations. Lastly, I want you to learn about the basics of amateur astronomy.

First, we will discuss some of the most basic concepts in astronomy (Part 1). For the most part, this knowledge will not directly apply to any of the observations we will learn to make. Despite that, I do believe that every amateur telescope user should at least possess some of this knowledge. Secondly, we will dive into some of the science and mathematics behind how telescopes work (Part 2). Before you use any scientific tool, you should know how it works. Afterwards, we will finally start discussing what to observe and how to observe it (Part 3).

After you read and understand the information presented in this book, you should be ready to find more observable astronomical targets on you own. The targets presented in this book are some of the easiest and most rewarding targets you can visually observe or image with a small telescope.

Thank you for purchasing this book. I hope you enjoy it.

Part 1: Basic Astronomy

Before we start talking about how and when to observe objects with a telescope, we need to understand how some of the science works. We will only discuss the absolute basics of astronomy. Astrophysics will not be considered. If you like Part 1 of this book, you may want to consider purchasing other books on these topics. This material is presented in such a way that you do not need a scientific background to understand it. A background in science, however, will make the material easier to understand.

Astrology is not *astronomy*. If you bought this book hoping to learn about horoscopes, you will be dissapointed.

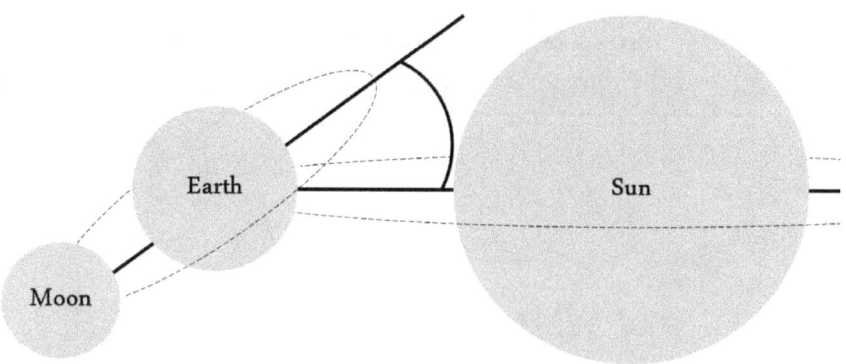

The above image shows (actually, exaggerates) the tilt of the moon's orbit around the Earth, with respect to the Earth's orbit around the sun. We'll be talking about this angle during this chapter so burn this image into your memory!

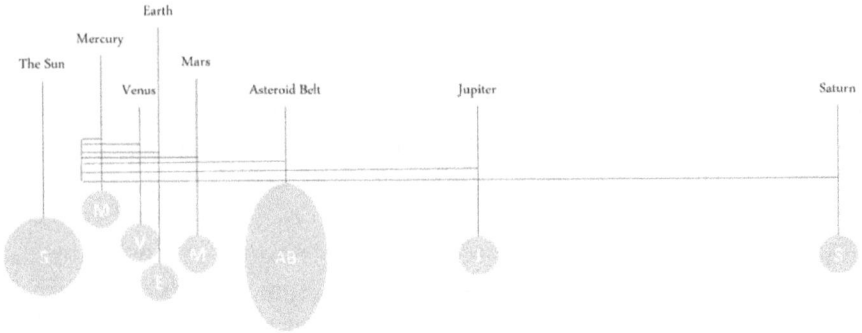

The scale of our solar system is not very intuitive. While the sizes of these bodies are not drawn to scale, the distances between them are drawn roughly to scale.

Object(s)	Distance to Sun (AU)	Distance to Earth (AU) (Closest Point)	Distance to Earth (AU) (Furthest Point)
Our Solar System			
Sun	0	1	1
Mercury	0.39	0.61	1.39
Venus	0.72	0.28	1.72
Earth	1		
Mars	1.52	0.52	2.52
Asteroid Belt	2.7	1.7	3.7
Jupiter	5.2	4.2	6.2
Saturn	9.54	8.54	10.54
Uranus	19.18	18.18	20.18
Neptune	30.06	29.06	31.06
Kupier Belt	50	49	51

Chapter 1

The Sun, Moon, and Earth

We live on the Earth. The Earth is a remarkable planet for many reasons. One of them is the fact that it is able to sustain life that reads books like this one.

We (human beings) need to breathe diatomic ("two atoms") oxygen, O_2, to survive, but only about 21% of the air is diatomic oxygen. At least ~18% of the air needs to be O_2 for us to breathe, but if the levels of O_2 were to exceed ~21%, then all organic material (like you and I) would be flammable (Nooo!). We exhale carbon dioxide, CO_2 (the classic greenhouse gas), but only around 0.04% of the air is CO_2. This begs the question, what else is in the air?

Nearly 78% of the air is the inert gas, diatomic nitrogen, N_2. Nitrogen is a non-reactive gas but if the concentration of nitrogen in the air was to ever exceed slightly higher than 80%, all human beings would die of asphyxiation (since it would exclude O_2 from our lungs). The composition of air is a delicate balance. Nearly 1% is the noble gas, Argon, and the rest contains very trace amounts of Neon (Ne), Helium (He), Methane (CH_4), diatomic Hydrogen (H_2), water vapor (H_2O), and the noble gas, Krypton (Kr).

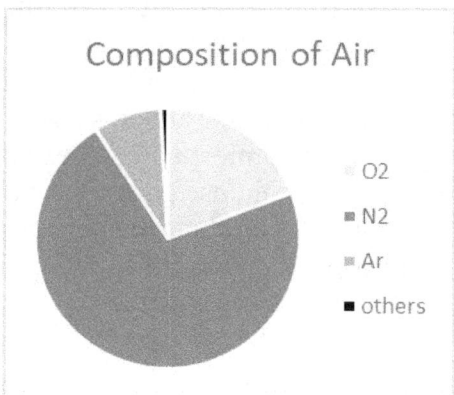

Plants "breathe" in CO_2 and "exhale" O_2 during photosynthesis, so it is probably a good idea we keep them around. More scientifically, plants "take in" carbon dioxide and water, and the light from the sun supplies the energy required to convert them into sugar and oxygen. The sugar produced can be used by the plants as a source of energy (like food to humans).

$$6CO_2 + 6H_2O + \text{(energy from sunlight)} \rightarrow C_6H_{12}O_6 + 6O_2$$

Photosynthesis is estimated to have begun on Earth around 3 billion years ago. The first oxygen molecules created by this process were primarily used to oxidize all the iron laying around (when iron is oxidized, it rusts). The oxygen continued to do this until it built up in the atmosphere. About 500 million years ago, the first land plants started to grow.

The Earth "breathes" too. In the fall, the trees and plants start to die, so the air becomes slightly concentrated in CO_2 (since there is less plant life around to take it in). In the spring, the trees and plants start to come to back to life, so the air becomes slightly less concentrated in CO_2 (since there is more plant life around to take it in and convert it into O_2). Throughout a year, the levels of CO_2 in the atmosphere fluctuate as the Earth takes these season-long "breaths".

A lot of people are worried about the amount of CO_2 in our atmosphere (0.04% may seem small, but it can still have an effect). CO_2 allows the sun's radiation to penetrate the atmosphere and warm up our planet, but it does not allow the resulting heat to escape. This is the greenhouse effect and it is a good thing. We need *some* greenhouse gases in our atmosphere to keep our planet warm. If the atmosphere had no greenhouse gases, we would all freeze to death. Obviously, we need *some* but not *too much* and it is this delicate balance that people are worried about.

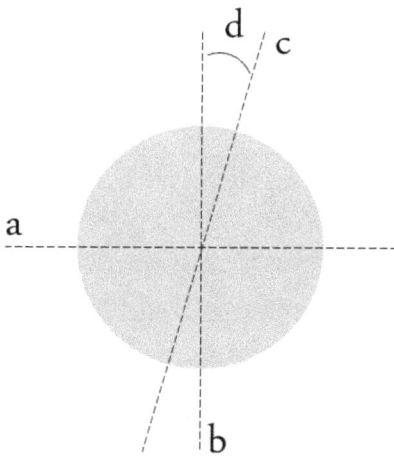

Figure 1.1 – The Earth and its axis. a is the plane of the orbit of the Earth around the sun, b is the axis perpendicular to a, c is the axis of rotation of the Earth, and d is the angle between c and b.

The Earth orbits the sun, but the Earth-sun system is much more complicated than that. The Earth spins on an axis and that axis is tilted at an angle of ~23.5 degrees with respect to the axis perpendicular to the Earth's orbit around the sun (figure 1.1).

The Earth completes a spin around its tilted axis every 24 hours (1 day). The Earth completes one orbit around the sun every year. It should be obvious, now, that humanity defined a day to be the amount of time it takes the Earth to complete one spin and defined a year to be the amount of time it takes the Earth to complete one orbit around the sun. What they chose to define a month is slightly more complicated.

The Earth orbits the sun, but the Earth has a little buddy that follows it during its orbit: *the moon*. The moon completes one orbit around the Earth every ~29.5 days, or approximately once a month. It actually only takes ~27 days for the moon to complete an orbit, but it spends the rest of the time "catching up" to the additional distance the Earth has traveled during this time.

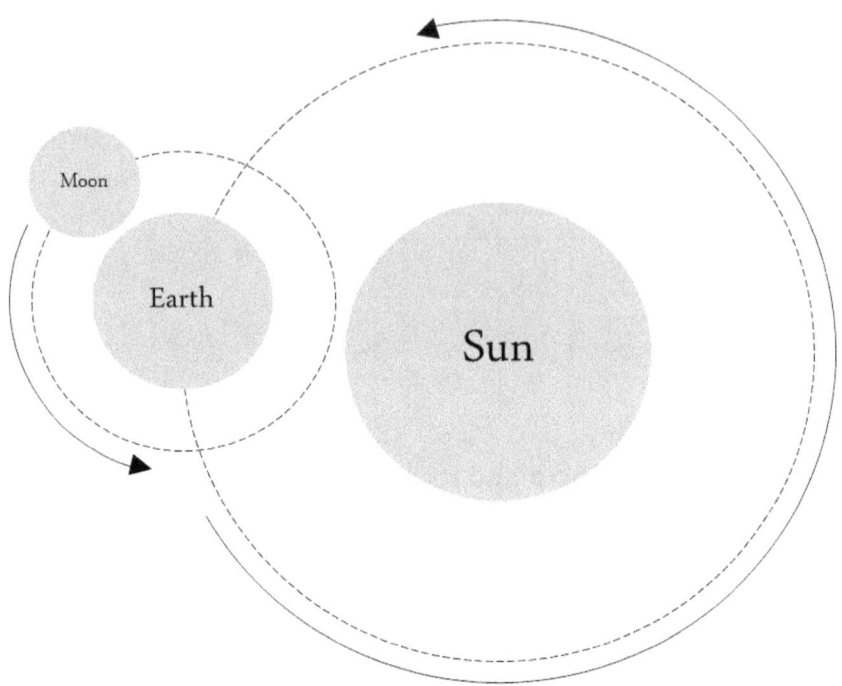

Figure 1.2 – The Earth's and the moon's respective orbits.

The word month comes from the word "moon". The moon's orbit around the earth is tilted by ~5 degrees with respect to the orbit of the Earth around the sun, but this angle is variable (it changes over time). The variation in the moon's tilted orbit explains the occurrence of lunar and solar eclipses. Normally, when we see the full moon, the moon is tilted away from the Earth-sun orbit so that sunlight can bounce off the moon and into our eyes (see figure 1.3 on the next page).

The following is important:
The moon has no internal, independent source of light. The moon is only visible because light from the sun bounces off the moon and into our eyes. The same is true for the planets. In contrast, stars do have an internal source of light. Stars are like light bulbs and planets are like paintings in a dark room. You can only see the paintings when the lights are turned on!

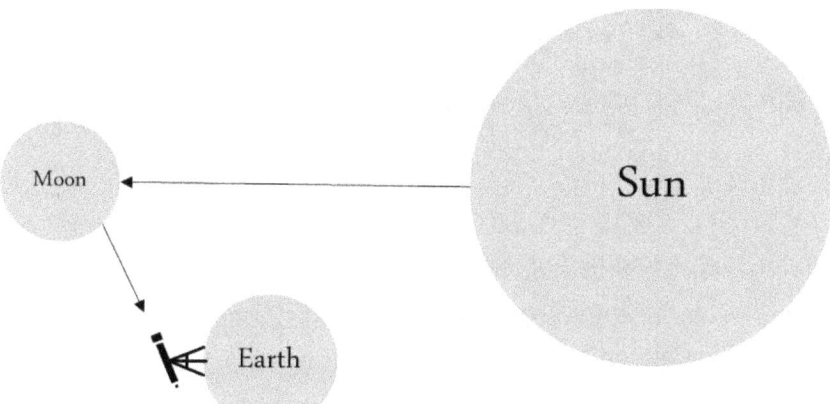

Figure 1.3 – When we see the full moon, the moon is tilted away from the plane of the Earth-sun orbit. This allows sunlight to bounce off the moon and onto the Earth.

Sometimes, however, the moon may be in-line with the Earth-sun system (the angle between the moon's orbit around the Earth and the Earth's orbit around the sun is zero). If the moon is located between the sun and the Earth (at the new moon position), a solar eclipse occurs (A in figure 1.4). The moon blocks part of the sun, and therefore you cannot look at solar eclipse with the naked eye (you are still looking at the sun, just a smaller fraction of it). If the moon is located past the Earth (at the full moon position), a lunar eclipse occurs (B in figure 1.4). This is perfectly safe to look at because it is no more dangerous than looking at the moon (which you are, obviously, allowed to do).

We will learn more about the lunar phases in chapter 8, but lunar eclipses can only occur during a full moon and solar eclipses can only occur during a new moon. Interestingly enough, a solar eclipse that occurred in 1868 was used to prove the existence of Helium. Pierre J. C. Janssen analyzed the light coming from the corona of that solar eclipse and found a line in the spectrum at 587.5 nanometers. This proved the existence of a new element, Helium.

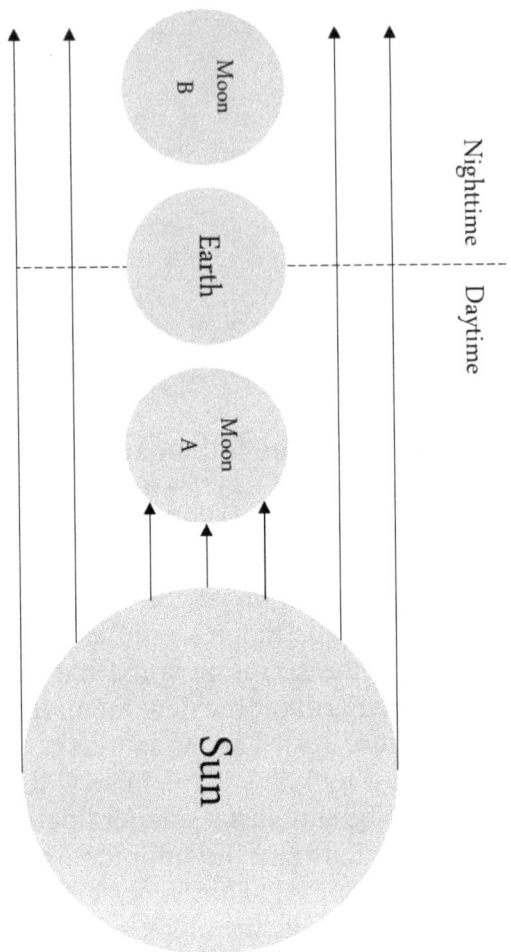

Figure 1.4 – Solar (A) and lunar (B) eclipses.

The moon may appear to be red during a lunar eclipse and this can be attributed to Earth's atmosphere. A lunar eclipse occurs when the Earth blocks sunlight from reaching the moon. According to this logic, no sunlight should be able to reach the moon so the moon should not be visible during a lunar eclipse (since no light can bounce off it). The moon may appear red during a lunar eclipse because the Earth's atmosphere refracts and bends red-orange light towards the moon and blue light out into the rest of the universe.

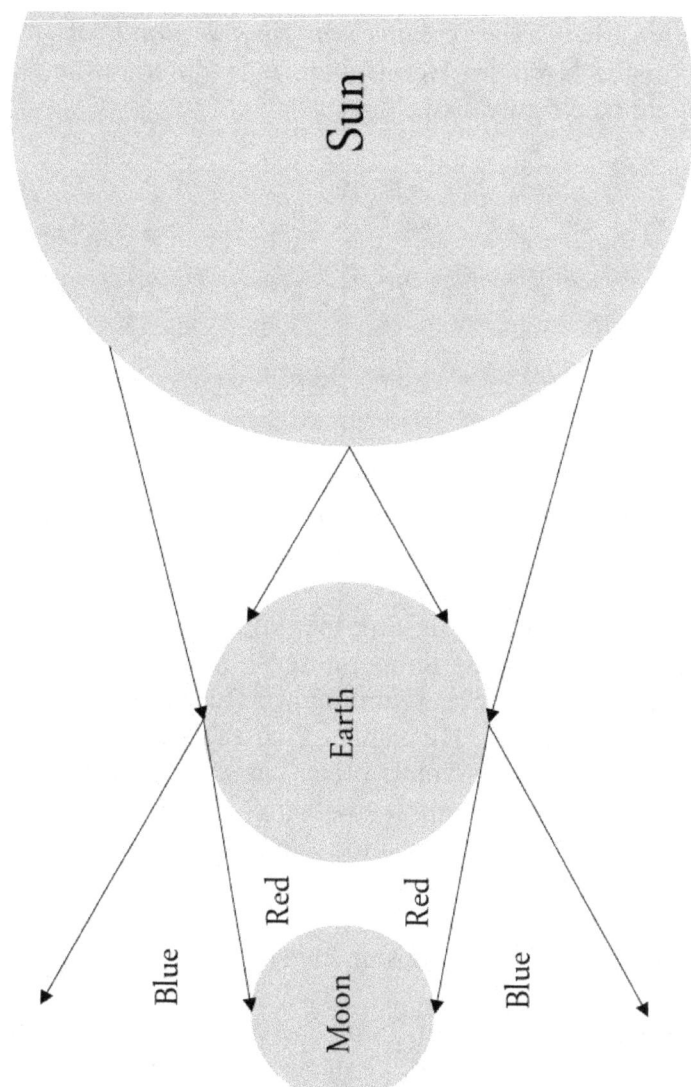

Figure 1.5 – The red ("blood") lunar eclipse moon.

This red light can then bounce off the moon and into our eyes. The region of red light is called the "umbra" and the region of blue light is called the "penumbra". In summary: we can credit the appearance of the red moon during a lunar eclipse to the Earth's atmosphere.

The Earth's tilted axis explains why, in the month of June, it is summer in the Northern Hemisphere but winter in the Southern Hemisphere (see figure 1.6).

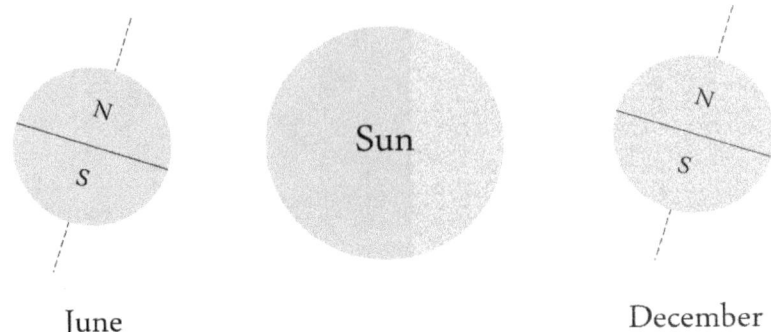

June   December

Figure 1.6 – The seasons are different in the Northern and Southern Hemispheres.

In June, more sunlight is hitting the Northern Hemisphere. More sunlight means higher temperatures. In June, the Northern Hemisphere is experiencing summer. In December, more sunlight is hitting the Southern Hemisphere, so the people there will be experiencing summer. The United States of America is in the Northern hemisphere (along with other countries). That is why this book only considers observations in the Northern hemisphere (most of the readers live there).

Now we can start discussing some basic astronomy.

You may have noticed that, in all the drawings so far, the Earth, sun, and the moon are all similar in size. You may have also noticed that the orbits of the moon around the Earth and the Earth around the sun are circular. Both are incorrect, but these simplifications make creating the images much easier.

Table 1.1 lists the approximate diameters of various objects in our solar system.

Object	Diameter (miles)
Sun	864,576
Earth	7,917
Moon	2,159
Mars	4,212
Jupiter	86,881
Saturn	72,367

Table 1.1 – The approximate diameters of several objects in our solar system

Table 1.1 provides strong evidence that the images you have seen thus far have been scientifically inaccurate (but are still adequate for getting the point across).

Now that we know the relative sizes of these objects, how far exactly are they from one another? In order to answer this question accurately, we need to know that moon's orbit around the Earth and the Earth's orbit around the sun are not circular (but they are close to circular). In fact, these objects orbit in ellipses. All of the planets orbit the sun in an elliptical shape.

When the moon is at is closest point to the Earth (perigee), it is approximately 225,000 miles away. When the moon is at is farthest point away from Earth (apogee), it is approximately 252,000 miles away. We know that the diameter of the Earth is almost 8,000 miles (table 1). When the moon is at its closest point to Earth, you can still fit around 28 Earths between the Earth and the moon. Figure 1.7 shows this roughly to scale.

Figure 1.7 – The distance between the moon and the Earth drawn approximately to scale. Relative sizes also drawn roughly to scale.

It should be obvious why I neglected to draw those previous images to scale – there would not be enough room.

The distance between the Earth and the sun is even more ridiculous. Amateur astronomers are not generally interested in the actual distance between the Earth and the Sun at different points in Earth's orbit. They are, however, very interested in the radius of the "circular" orbit between the Earth and the sun. I know. I just said that the Earth's orbit around the sun is not circular but using the radius of the "pseudo-circular" orbit of the Earth around the sun is quite a useful trick that astronomers have made standard practice.

This pseudo-circular radius is *roughly* equal to the distance between the Earth and the sun and it is defined as 149,597,871,000 meters, or 93 million miles (11,625 Earths). This radius defines the **astronomical unit** and that is in bold for a reason. The astronomical unit is used as a convenient way of measuring distances between objects in our solar system. The Earth and the Sun are 1 astronomical unit (AU) apart – see figure 1.8.

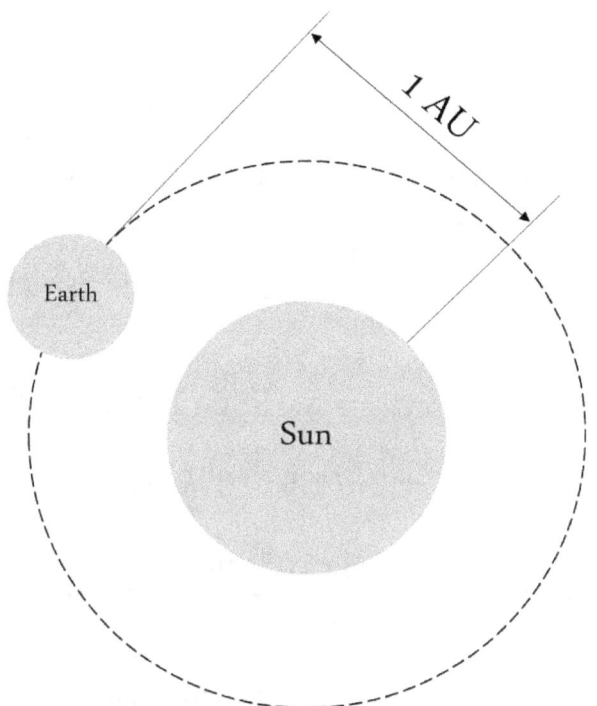

Figure 1.8 – The definition of the astronomical unit

Table 1.2 lists the distances between the planets and the sun in terms of the astronomical unit.

Planet	Distance from Sun
Mercury	0.39 AU
Venus	0.72 AU
Earth	1 AU
Mars	1.52 AU
Jupiter	5.20 AU
Saturn	9.54 AU
Uranus	19.18 AU
Neptune	30.06 AU

Table 1.2 – Distances in AU

We will talk more about orbits in the next chapter. It is there where you will see how practical using the AU is. If you are, however, very eager to use mathematics (which most of you probably are not), we can calculate how fast the Earth (or any planet) is moving during its orbit around the sun using basic physics. Velocity is equal to the distance traveled by an object divided by the amount of time it took the object to travel that distance. In one year, the Earth travels the distance of the circumference of its "pseudo-circular" orbit.

$$time = 1year = 3.1536x10^7 seconds$$

$$distance = 2\pi r = 2\pi(1.5x10^8 km) = 9.4x10^8 km$$

$$veloctiy = \frac{(9.4x10^8 km)}{(3.1536x10^7 sec)} = 30\ km/s$$

The Earth travels around the sun at approximately 30,000 meters per second (67,000 mph). It should be noted that the orbital period, P, of a planet is equal to the distance traveled by the planet in one orbit around the sun divided by the velocity of that planet during its orbit.

$$P = \frac{(2\pi r)}{v} = \frac{distance}{(distance/time)}$$

The moon is a particularly easy and fun object to observe with any telescope. More information about the moon can be found in chapter 8. It is there where we will discuss the lunar phases and other fun moon stuff.

Chapter 2

Planets & Planetary Orbits

Kepler's Laws

Kepler's three laws of planetary motion give us a more complete picture of how the planets orbit the sun.

Kepler's first law states that a planet moves around the sun in an elliptical orbit. An elliptical has two foci, and the sum of the distances from these two foci to any point on the elliptical is always the same. Kepler's first law also states that the sun is always one of these foci.

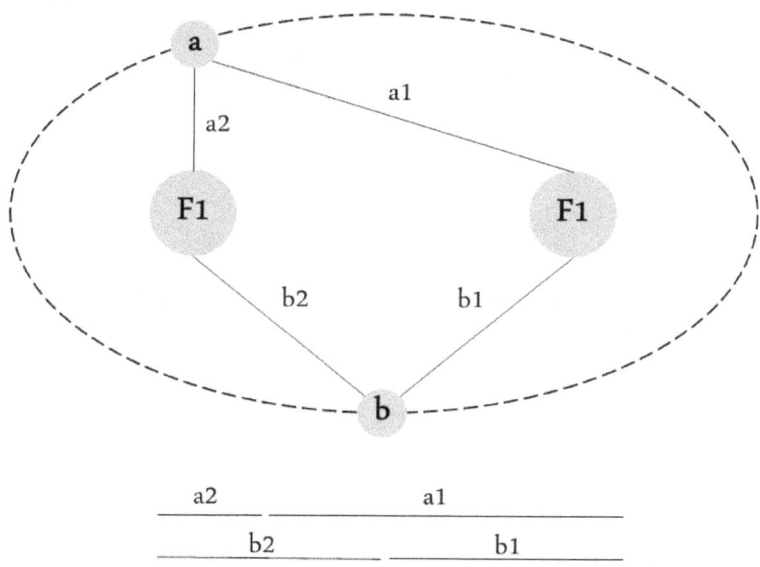

Figure 2.1 – F1 and F2 are the foci of this planet's elliptical orbit. a1, a2, b1, and b2 are the distances from the foci to a point on the orbit. "a" and "b" are points on the elliptical.

$$a_1 + a_2 = X = b_1 + b_2$$

Kepler's second law is a little more difficult to visualize. First, it is important to note that a planet travels faster when it is closer to the sun. During the same amount of time, *t*, a planet will travel a greater distance when it is closer to the sun. Kepler's second law states that the radius "sweeps out" in equal areas during the same amount of time, *t*, at any point in the planet's orbit. Figure 2.2 may be more helpful than words. In figure 2.2, A1 and A2 are equal since the same amount of time, *t*, has passed in both situations.

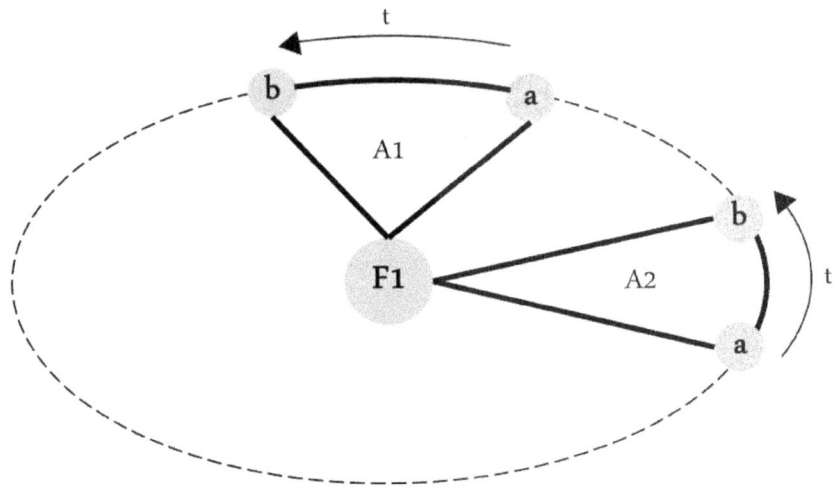

Figure 2.2 – Kepler's second law states that A1 and A2 must be equal since the time that passed by, t, was the same in both orbital segments.

$$A_1 = A_2$$

Kepler's third law is more of an equation.

$$P^2 = d^3$$

Where P is the orbital period of the planet (in Earth years) and d is the distance from the planet to the sun (in terms of the astronomical unit). This equation is remarkably accurate to what is observed. Table 2.1 lists the distances between the planets and the sun (d) and the orbital period of the planets around the sun (P).

Planet	Distance from Sun (AU)	Orbital Period (Earth-years)
Mercury	0.39	0.24
Venus	0.72	0.62
Earth	1	1.00
Mars	1.52	1.88
Jupiter	5.20	11.86
Saturn	9.54	29.46
Uranus	19.18	84.01
Neptune	30.06	164.80

Table 2.1 – Data for Kepler's third law.

Table 2.1 lists data that we observe, so let's see just how accurate Kepler's third law is.

EXAMPLE – Calculate the orbital period of Mercury using the distance from the sun in table 3.

$$P = \sqrt{(0.39^3)} = 0.244 years$$

EXAMPLE – Calculate the orbital period of Jupiter using the distance from the sun in table 3.

$$P = \sqrt{(5.20^3)} = 11.858 years$$

EXAMPLE – Calculate the distance between the sun and Saturn using the orbital period in table 3.

$$d = (29.46^2)^{(1/3)} = 9.538 AU$$

It appears that Kepler's third law is valid since the results are consistent with the data in table 3.

The Planets

Planets further from the sun than the Earth are called superior planets and planets closer to the sun than the Earth are called inferior planets.

Mercury is the smallest planet and the closest planet to the sun in our solar system. Its axis is not tilted, and it is the only planet with this trait. It takes approximately 59 Earth days for Mercury to complete one spin around its axis. Mercury has no moons. Mercury is not easily observed with the small telescope, so no further discussion is necessary in this book. More advanced solar observers & imagers may enjoy mercury during rare solar transits.

Venus, on the other hand, is very interesting. Venus is Earth's planetary neighbor that is closer to the sun. Venus' atmosphere is highly concentrated in sulfuric acid and CO_2. As we learned in chapter 1, CO_2 is a greenhouse gas. It allows the sun's light to pass through it, but it traps the heat that tries to escape.

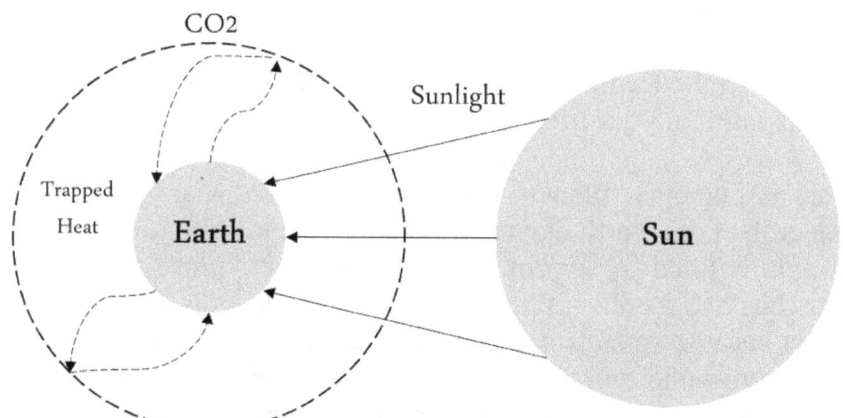

Figure 2.3 – Carbon Dioxide is a greenhouse gas.

Venus is a friendly reminder of just how ugly this effect can get. Venus' average surface temperature is higher than Mercury's even though Mercury is closer to the sun. You could argue that this is an unfair comparison since Mercury has no atmosphere, but it is still interesting to see the impact of greenhouse gases. It should be noted that CO_2 is not the only greenhouse gas. Water vapor, for example, is a greenhouse gas on Earth since there is always some water vapor in the air. Venus is so bright because the sulfuric acid clouds in its atmosphere are very reflective. Scientists had actually proposed that sulfuric acid be sprayed into the upper atmosphere of the Earth to reflect some of the sunlight away and reduce the amount of light that heats up the Earth, but fears about acid rain ended the research (as far as I know).

Venus has no moons and it takes approximately 243 Earth days for Venus to rotate once about its axis. Venus spins in the opposite direction to the way it orbits the sun. This is called retrograde rotation and only Venus and Uranus rotate this way. Venus' axis is tilted 177 degrees (it is almost completely upside down). Venus has some unique properties, and it is sometimes visible in the night or morning skies.

We live on the Earth. You should know some information about it by this point in your life. Take care of it.

Mars is our other planetary neighbor. It takes Mars just a little longer than one Earth day to complete a rotation about its axis and its axis is tilted by approximately 24 degrees (close to Earth's). Mars has two moons – Phobos and Deimos. Mars is an interesting planet because scientists hope to get humans to live on it someday. Mars' atmosphere is mostly composed of carbon dioxide (96%), but Nitrogen and Argon each account for approximately 2% of its composition. Trace amounts of water, carbon monoxide, methane, and similar gases are also present.

Jupiter and Saturn are the giants of our solar system. Both Saturn and Jupiter complete one rotation about their axis in just a little less than half an Earth day. Jupiter spins so fast around its axis that its equator bulges out. Jupiter has more than 75 moons, Saturn has 62 moons (as of 2021), and both Saturn and Jupiter have rings. A probe in 1995 detected that Jupiter was composed of nearly 98% Hydrogen and Helium.

Four of Jupiter's moons are typically visible with a small telescope: Callisto, Europe, Ganymede, and Io (they are called the Galilean Satellites in honor of the man who discovered them). These moons are visible using a small telescope because they are the largest. Europa is similar in size to our own moon; and Callisto and Ganymede are both slightly larger than Mercury. Io is the most volcanically active site in our solar system. It appears to be slightly yellow in color because of all the sulfur that results from this volcanic activity. The other three moons are made primarily of ice (Ganymede and Callisto are not as pure as Europa).

Saturn's rings are famous. If you have a decent telescope, you can see them for yourself (and you are going to want to try).

Neptune and Uranus are beyond the scope of this book, as they are generally not worth the trouble to observe with a small telescope, but you can make out a colored dot sometimes.

As far as small telescope observations are concerned: Venus, Mars, Jupiter, and Saturn are the planets most worthy of your money, time, and effort. These observations are discussed in chapter 9.

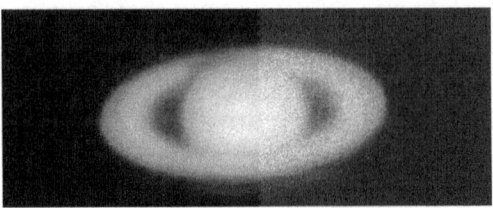

Saturn during poor seeing conditions. Basic rings visible using small telescopes.

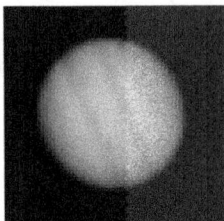

Jupiter. Cloud bands visible using small telescopes.

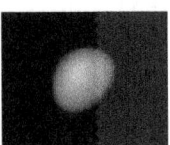

Venus. Phases visible using small telescopes.

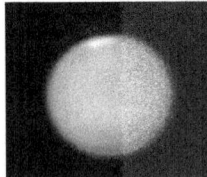

Mars. Polar caps and minor surface detail may be visible.

Chapter 3

Light and the Stars

Light is an electromagnetic wave that exhibits wave-particle duality (it behaves like a particle in some situations and like a wave in other situations). There is a fundamental equation that relates the frequency and wavelength of light, $v = c\,\lambda^{-1}$, where v is the frequency of the light, c is the speed of light (a constant), and λ is the wavelength of the light. This equation tells us that as the wavelength of light increases, the frequency decreases. In other words, they are inversely proportional.

Light in the visible spectrum (the light we can see) has wavelengths in the range of 400-700 nanometers (nm) (10^{-9} meters). Blue/Violet light has wavelengths of ~400 nm and red light has wavelengths of ~700 nm. The visible spectrum can be remembered as ROYGBIV (pronounced *Roy G Biv*) (Red, Orange, Yellow, Green, Blue, Indigo, and Violet).

Color	Wavelength (nm)
Red	~ 610-700
Orange	~ 590-610
Yellow	~ 570-590
Green	~ 500-570
Blue	~ 450-500
Violet	~ 400-450

Light outside of these boundaries exists, but the human eye cannot interpret it. UV (ultraviolet) light has wavelengths shorter than visible light all the way down to 10 nm. IR (infrared) light has wavelengths longer than visible light all the way up to 1 millimeter (mm) (10^{-3} meters).

Microwaves and radio waves have wavelengths longer than IR light; X rays and gamma rays have wavelengths shorter than UV light (cosmic rays have wavelengths even shorter than gamma rays).

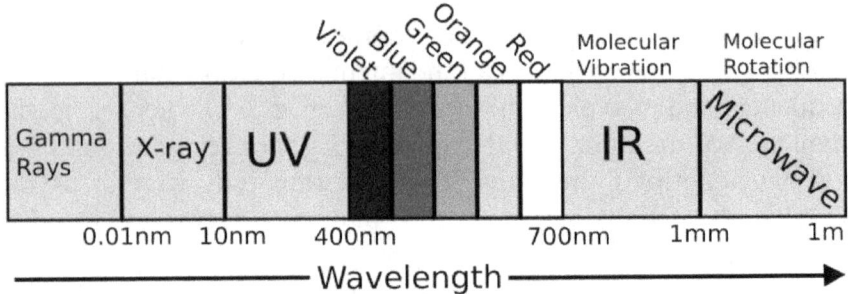

Figure 3.1 – Types of electromagnetic radiation (light).

Molecules vibrate in IR light and rotate in microwave light. This is how microwave ovens and IR scans work. Microwave ovens emit (you guessed it) microwave light, or microwave radiation, at a certain wavelength (or frequency) that rotates water molecules. The rotation of these water molecules causes friction and the resultant heat cooks whatever is inside the microwave.

Chemists use IR scans to analyze the structure of chemical compounds. IR scans can reveal if certain groups of chemicals (called functional groups) are present. A typical IR scan of a chemical compound has "peaks" and these peaks correspond to certain functional groups. Some functional groups absorb unique wavelengths of IR light and vibrate. For example, if a compound contains an alcohol (oxygen-hydrogen) group, there will be a broad, strong peak on the IR scan from ~3100-3500 cm[-1] (wavenumbers).

When an object appears to be a certain color, it is because the object's chemical surface reflects the wavelength of visible light associated with that color and absorbs the others.

Color Absorbed	Wavelengths Absorbed	Object Color
Red	~ 650-700 nm	Green
Orange	~ 590-650 nm	Green-Blue
Yellow	~ 570 – 590 nm	Blue
Green	~ 490 – 570 nm	Red
Blue	~ 420 – 490 nm	Yellow
Violet	~ 400 – 420 nm	Yellow-Green

An UV/VIS scan can be found below. An UV/VIS scan shows what wavelengths of UV and visible light a chemical species absorbs.

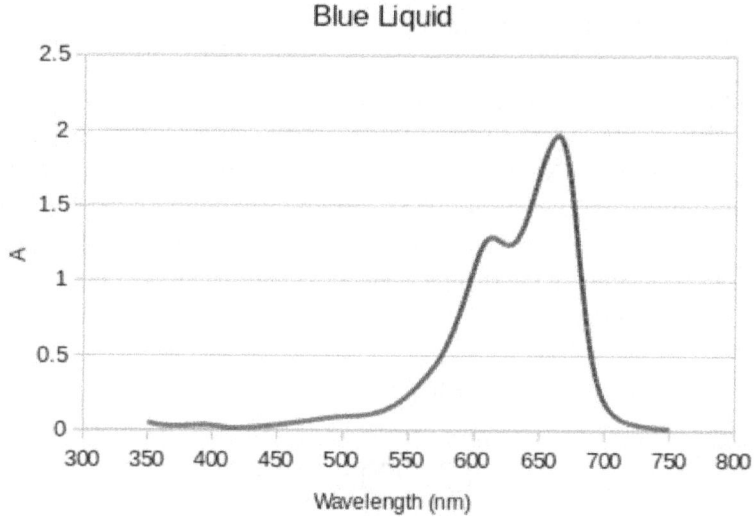

Figure 3.2 – UV/VIS spectrum of a blue liquid. Blue light is reflected, while red light is absorbed.

In this UV/VIS scan, we can clearly see that the liquid being tested absorbed wavelengths of light from 600-670nm and reflected wavelengths of light from 400-500nm. The liquid was blue because it reflected light near the blue end of the spectrum.

The sun's radiation closely resembles that of a blackbody with a temperature of 5,800 Kelvin (the Kelvin scale is just another temperature scale like degrees Celsius or degrees Fahrenheit). Indeed, the surface temperature of the sun is approximately 5,800K. At ground level (after passing through the atmosphere), approximately 5% of the sun's light is UV light, 45% is visible light, and 50% is infrared light. Scientists classify the UV light that comes from the sun into three different sub-categories: UVA, UVB, and UVC. Ultraviolet C (UVC) light has wavelengths of 100-280 nm and most of this light gets absorbed in the Earth's atmosphere. Ultraviolet B (UVB) light has wavelengths of 280-315 nm and most of this light also gets absorbed in the Earth's atmosphere. UVC and UVB light cause the photochemical reaction in the atmosphere that converts O_2 into O_3 (ozone). UVC and UVB light are responsible for the ozone layer.

The ozone layer is more of a cycle between diatomic oxygen and ozone. First, sunlight with a wavelength of 100-240 nm (UVC) strikes O_2 and converts it into two oxygen radicals (extremely reactive chemical species). These oxygen radicals then react with other O_2 molecules and form ozone, O_3. Ozone absorbs sunlight with wavelengths of 250-315 nm (UVC and UVB) and this light converts it into O_2. This is a continuing cycle.

O_2 + (UVC light from the sun) $\rightarrow$ 2 O•
O• + O_2 $\rightarrow$ O_3
$2O_3$ + (UVC/UVB light from the sun) $\rightarrow$ $3O_2$

Older refrigerants, such as CFCs (chloro fluoro hydrocarbons) (R11 and R12 are two examples), interfere with this cycle when they are released into the atmosphere and, therefore, allow more UVB and UVC light to pass through what are called ozone holes. These refrigerants are no longer used for this reason.

UVB and UVC light are particularly harmful to DNA and will cause sunburn (this is why ozone holes are bad). That being said, you still need to be exposed to some of this light so that your body can synthesize vitamins D_2 and D_3 (the combination of these two is often referred to as vitamin D). The sun's light is simply electromagnetic waves and it contains no chemicals itself, but the energy from those UVB and UVC waves starts chemical reactions in your skin that creates vitamin D.

You do not need to be exposed to too much of this light. Your body only requires small amounts of vitamin D and can manage to make its own with little UVB and UVC exposure.

Ultraviolet A (UVA) light has wavelengths of 315-400 nm. This light is used in tanning beds because it was once thought to be safer than UVB and UVC light. Now we know that this is not the case, but people continue to tan in these beds. UVA light also damages DNA (hence the tan) but it does so indirectly by forming chemical species that do the damage. That being said, if you want to be tan you will likely and willingly expose yourself to this dangerous radiation no matter what I tell you.

We can assume that the sun supplies around 800 Watts per square meter (W/m^2) of power to the surface of the Earth on a normal, sunny day. This value changes depending on the position of the sun in the sky, the clarity of the sky, and the positions of the sun and the Earth. If we assume the value is 800 W/m^2, then we can say that, per square meter, we can get 400 Watts of power from the IR light, 360 Watts of power from the visible light, and 40 Watts of power from the UV light that hits the Earth. When scientists work on obtaining renewable energy from the sun's light, they generally work on getting the energy from the IR or visible light wavelengths since most of the power comes from these types of light.

Light travels at a constant speed of 2.99792558×10^8 meters per second. Nothing can travel faster than the speed of light and no object can travel at the speed of light. Despite how fast light travels, it still takes time to move from one place to another.

The moon is approximately 225,000 miles away at its closest point to Earth. In meters, that is 362,102,400. The time, t, something takes to travel a distance is equal to the distance traveled, d, divided by the velocity, v, in which that something traveled at. When the moon is closest to Earth, the light that bounces off the moon travels at the speed of light and it travels 362,102,400 meters before it gets to Earth and hits your eyes.

$$t = \frac{d}{v} = \frac{(362,102,400m)}{(299,792,558ms^{-1})} = 1.21 seconds$$

When the moon is at its closest point to the Earth, the light takes 1.21 seconds to get to us. When you look at the moon, you are looking at it how it was just over 1 second in the past. The moon is said to be around 1.3 light-seconds away (averaging its closest and farthest points). The sun is even farther away. The light from the sun takes a little longer than 8 minutes to get to us. The sun is said to be around 8.3 light-minutes away. The closest stars are 4 light-years away. When we look at a star, we are looking at how it was years in the past. The Andromeda galaxy is the closest major galaxy to our own, but it is still 2.5 million light-years away!

Sirius is the brightest star in the night sky. Polaris, the Northern star, is the 40-something[th] brightest star in the night sky. The brightness of a star (as observed from Earth) is called its apparent magnitude. Jupiter can have a magnitude of -2.5, and it is almost three times as bright as Sirius. Venus can have a magnitude of -4 at times. Saturn is typically as bright as a first magnitude star, but its magnitude can reach as low as 0.6. The lower the magnitude, the brighter it is. First magnitude stars have a magnitude of ~1, second magnitude stars have a magnitude of ~2, and so on. Light pollution makes it difficult to see stars and deep sky objects that have higher magnitudes. The brightness of an object in the night sky depends on its luminosity and the distance between the object and the observer (you). The luminosity, L, of an object is defined as the amount of energy emitted by the object per unit of time.

$$L = \frac{(EnergyEmitted(Joules))}{(time(seconds))} = Power(Watts)$$

The light an object emits can leave the object and travel in an infinite amount of directions. The combination of all these light pathways forms a sort of "light-sphere" (when all the pathways are given a finite length). The surface area of any sphere is equal to $4\pi r^2$, where r is the radius. In this case, the radius is equal to the distance between the central light source and the observer.

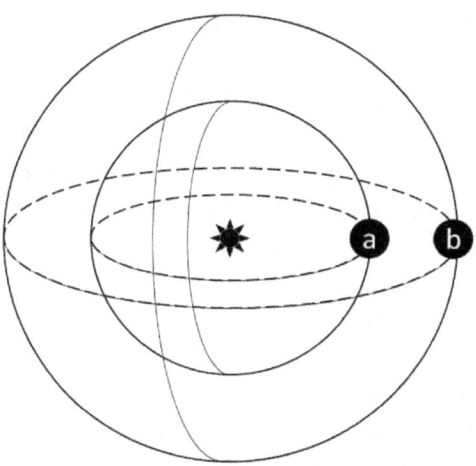

The surface area of this light-sphere gets larger as the distance between the observer and the object emitting the light increases. The surface area of the light-sphere with the observer at the "b" position would certainly be larger than the surface area of the light sphere with the observer at the "a" position. A fraction of this light will hit the observer. How bright does the object appear to be? Simply divide the luminosity of the object emitting the light by the surface area of the light-sphere with the radius of this light-sphere being the distance between the observer and the light.

$$Brightness = \frac{(L)}{(4\pi r^2)}$$

Where L is the luminosity of the object and r is the distance between the observer (you) and the object.

Calculate the brightness of the sun. The luminosity of the sun is roughly equal to 3.846 x 10^{26} Watts and the sun is roughly 149,597,871,000 meters away (watts and meters are used to keep the units consistent). This value is referred to as the solar constant.

$$B = \frac{(3.846x10^{26} J/s)}{(4\pi(149597871000meters)^2)} = 1370 \, W/m^2$$

Hydrogen fuels the stars and accounts for approximately 88% of the atoms in the universe. The presence of Hydrogen, Helium, and Lithium in the universe is used as evidence that the Big Bang occurred. More interestingly, the presence of Boron in the universe indicates that the Big Bang was an inhomogeneous event. Boron cannot be made within stars, so its presence indicates that it was probably made in neutron-rich regions during the Big Bang.

The center of some stars (like our sun) can have a temperature of nearly 15 million degrees centigrade. At this high of a temperature, nuclear fusion will occur. Nuclear fusion occurs when two atoms collide and their nuclei fuse together. A lot of energy is required for this process to initiate (hence the high temperatures), but even more is released after it occurs. First, two Hydrogen nuclei fuse and form Deuterium and release a positron. Then the Deuterium and another Hydrogen atom form Helium-3. When two Helium-3 atoms collide, normal Helium is formed and two hydrogens are ejected (and can undergo the process again). This is the basic process of the nuclear fusion that takes place in the stars.

(1) $2 \, ^1H \rightarrow \, ^2H + positron$
(2) $^2H + \, ^1H \rightarrow \, ^3He$
(3) $2 \, ^3He \rightarrow \, ^4He + 2 \, ^1H$

$3 \, ^1H$ go in but only $2 \, ^1H$ come out. That means, eventually, a star will run out of fuel. More on this later.

A tremendous amount of energy is released during this process. Iron is the heaviest element that can be made in the center of a star. Iron has the most tightly bound nucleus, so no more energy can be released if it fuses with other nuclei. Heavier elements are made in supernovas.

Scientists have worked on finding a way to harness the nuclear fusions that occur within the stars for quite some time now. Why? If scientists could figure out a way to perform these fusion reactions on Earth and figure out a way to recover this energy, we could power society indefinitely. When 4 grams of hydrogen converts into 4 grams of helium, 2.5 billion kilojoules of energy is released. This is equivalent to the energy produced from burning more than 50 tons of oil. This problem is not trivial due to the fact that the fusions require temperatures of millions of degrees. We cannot achieve such high temperatures on Earth's surface.

Scientists who work on this, want to force these fusions to occur at much lower temperatures. This is called "cold fusion".

In 1989 Martin Fleischmann and Stanley Pons reported that they had successfully completed something similar to cold fusion, but they gave little detail. Scientists continued to work on their ideas, but no one could replicate the results (it was later discovered that Fleischmann and Pons faked some data). As a result, the dream of cold fusion powering our society has since been dead. Not many people research cold fusion anymore since it is generally regarded as a non-sense science. If they do, they typically refer to it as low-energy nuclear reactions (LENR).

Each chemical element absorbs different wavelengths of light, and this is how scientists calculate the composition of stars and interstellar clouds. For example, most spectra of deep sky objects reveal that they contain large amounts of hydrogen.

Sometimes a star that appears to be a single star to the human eye is actually two or more stars that are just close together. A telescope (or even binoculars) can sometimes "split" these stars. When you "split" a star, you reveal that what appeared to be one star to the unaided eye, is really two stars. Sometimes these double and multiple star systems contain different colored stars.

"Nebula" used to be a term that referred to any distant astronomical object besides comets or planets. Today, it has quite a different definition.

A *galaxy* is a group of billions of stars that sometimes have their own planets, just like our sun. If you are interested in astronomy, you probably already know that the concept of the over-abundance of galaxies (and therefore planets) in our universe is used to defend the theory that the Earth is not the only planet that supports life. Since galaxies contain billions of stars, who can say that none of those *billions* of stars have a planet orbiting it that sustains life? If you disagree with this theory, then you certainly disagree with statistics. Hydrogen in essential for life on Earth (since water contains 2 hydrogen atoms). Since almost 90% of the atoms in the universe are hydrogen, it would be statistically impossible for all this hydrogen to create only one life-sustaining planet in the infinite universe. That being said, you are still free to believe whatever you wish. Statisticians and astrophysicists have no right to tell you what you ought to believe since no one actually knows any of this is true. If you are interested in this, I suggest reading about the "Drake equation" and the "Fermi paradox".

Globular clusters are groups of hundreds of gravitationally bound stars. M13 is an example of a globular cluster. *Open clusters* are groups of dozens of stars that are not gravitationally bound to each other. These stars will eventually separate over time. M44, the beehive cluster, is an example of an open cluster. *Emission nebula* are clouds of high temperature gases (they are usually red because they are composed of mostly Hydrogen) and *Reflection nebula* are clouds of dust that reflect the light of nearby stars (they are usually blue due to the efficient scattering of light).

Emission and reflection nebula are usually close to each other and can be referred to, collectively, as *diffuse nebula*. M42, the Orion nebula, is an example of a diffuse nebula. *Planetary nebula* have nothing to do with planets. They are simply shells of gas that a dying star gives off. M27, the dumbbell planetary nebula, is an example of a planetary nebula. A supernova occurs when a star dies and releases massive amounts of energy. *Supernovae remnants* are the leftover star pieces from a supernova. M1, the crab nebula, is a famous example.

The motions of the moon and sun are independent (for our purposes) from one another – the moon does not always rise when the sun sets (more on this later). But what about the planets, deep sky objects, and stars? The stars and deep sky objects move the same way through our night skies. That is why I use the constellations to help you find deep sky objects in chapter 9. For example, M41 is always going to be found approximately 4 degrees under alpha Canis Majoris (Sirius).

In the Northern Hemisphere, Polaris is the "north star". It is given such a name because it lies very close to the northern pole. The location of Polaris in *your* sky depends on *your* latitude. If you live near the North Pole, Polaris would be almost directly above you. Where I observe, the latitude is approximately 40 degrees, so Polaris can be found 40 degrees up from the horizon. If you have an equatorial mount and you live in the Northern Hemisphere, then you can consider Polaris to be the northern pole when you are setting up and aligning your stand.

Stars and deep sky objects move around the northern pole as the night progresses. If they are close to the pole, they appear to spin. If they are far away from the pole, they appear to rise in the east and set in the west, just like our sun and moon.

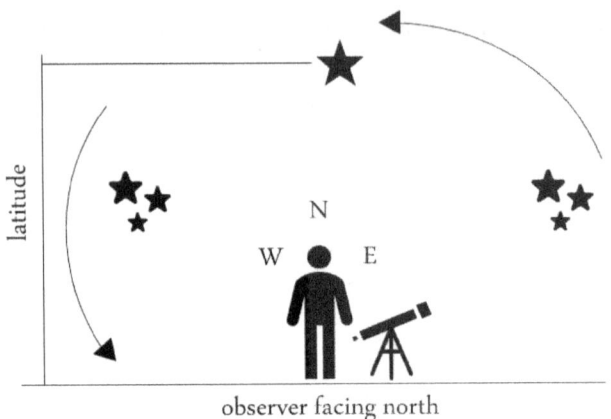

observer facing north

You may have noticed that many galaxies and nebulae have a name that follows the "M#" format. This is just one way to name a deep sky object. Objects named this way are called Messier objects and they are just a list of 110 galaxies and nebula Charles Messier thought could be confused for a comet. The numbers just represent the order in which he found them. The New General Catalogue (NGC) is a list of deep sky objects that was published in 1888 by the Royal Astronomical Society. The following table lists some examples (only objects with a common name are shown – not all deep sky objects have a common name).

Messier	NGC	Common Name
M42	1976	Orion Nebula
M44	2632	Beehive Cluster
M51	5194	Whirlpool Galaxy
M27	6853	Dumbbell Planetary Nebula
M1	1952	Crab Nebula
M8	6523	Lagoon Nebula
M57	6720	Ring Nebula
M20	6514	Trifid Nebula
M31	224	Andromeda Galaxy

Stars also have names. A long time ago astronomers decided that instead of pointing to a bright dot and referring to it as "that thing", they were going to name the stars. There are many ways to do this, but we will only discuss the simplest way. In chapter 9, the constellations are shown and some of the stars in those constellations are named using their common names. Common names are just simple names that give no information about the star.

Stars can also be named according to their magnitude and the constellation they are in (if they are in one). Consider a generic constellation A. In A the brightest star (lowest magnitude) would be called alpha A; the second brightest would be called beta A; and so on, following the Greek alphabet. The Greek alphabet can be found on the next page.

While this seems like a great system, there are many exceptions. Consider Ursa Major (the big dipper).

The magnitudes of the stars are in the parenthesis. It is clear that these stars are named from right to left with no consideration of their magnitudes. The alpha star is called *alpha Ursae Majoris*, the beta star is called *beta Ursae Majoris*, and so on.

Greek Alphabet (lowercase)

α	alpha
β	beta
γ	gamma
δ	delta
ε	epsilon
ζ	zeta
η	eta
θ	theta
ι	iota
κ	kappa
λ	lambda
μ	mu
ν	nu
ξ	xi
ο	omicron
π	pi
ρ	rho
σ	sigma
τ	tau
υ	upsilon
φ	phi
χ	chi
ψ	psi
ω	omega

Consider Ursa Minor (the little dipper).

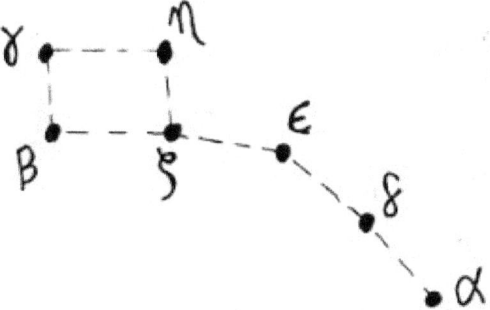

The alpha star in this constellation is called *alpha Ursae Minoris*, and so on. Alpha Ursae Minoris is the "North Star" and has the common name "Polaris". Now consider Leo.

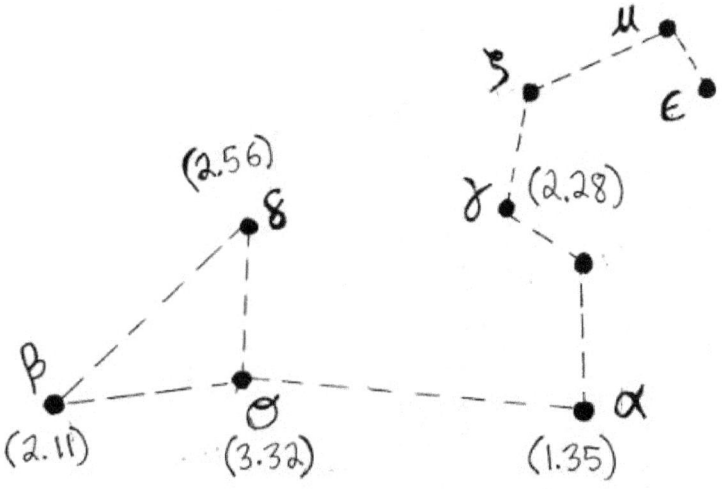

We can see that the stars in Leo roughly follow this naming rule. The alpha star in this constellation is called *alpha Leonis*, and so on. Although this naming rule has many exceptions, I do feel as though you should at least know it exists.

Some popular examples can be found on the following pages.

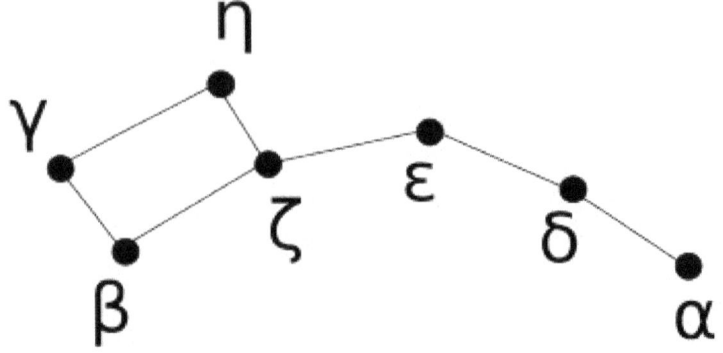

Greek letter + "Ursae Minoris"

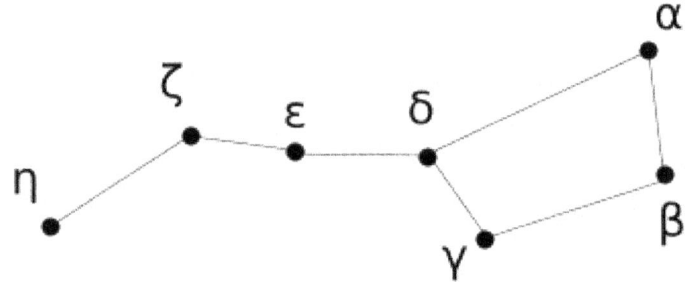

Greek letter + "Ursae Majoris"

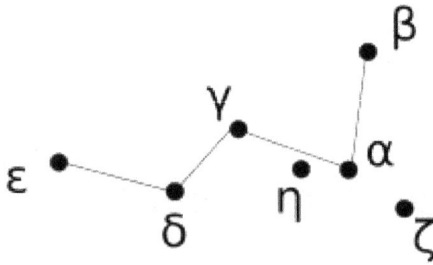

Greek letter + "Cassiopeiae"

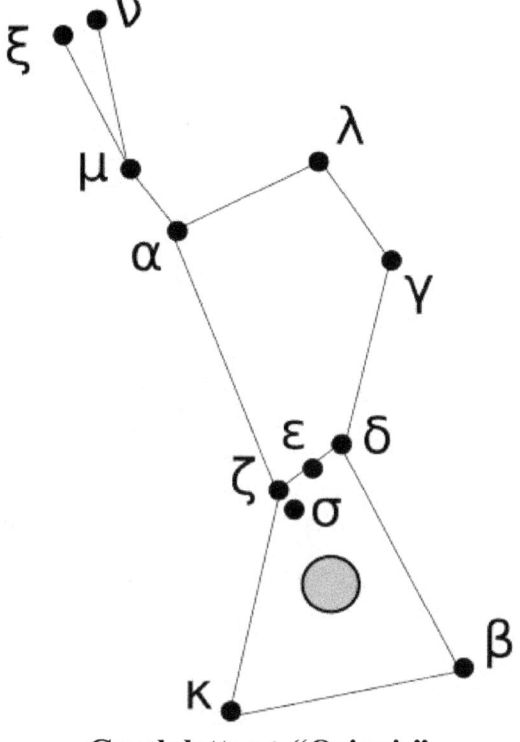

Greek letter + "Orionis"

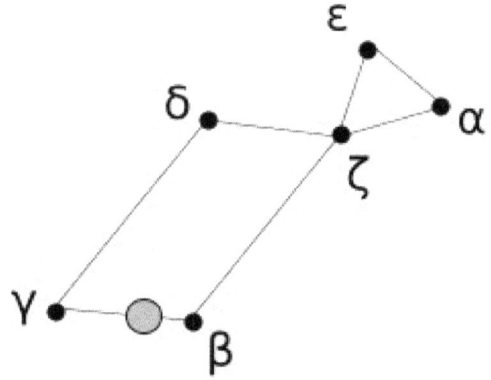

Greek letter + "Lyrae"

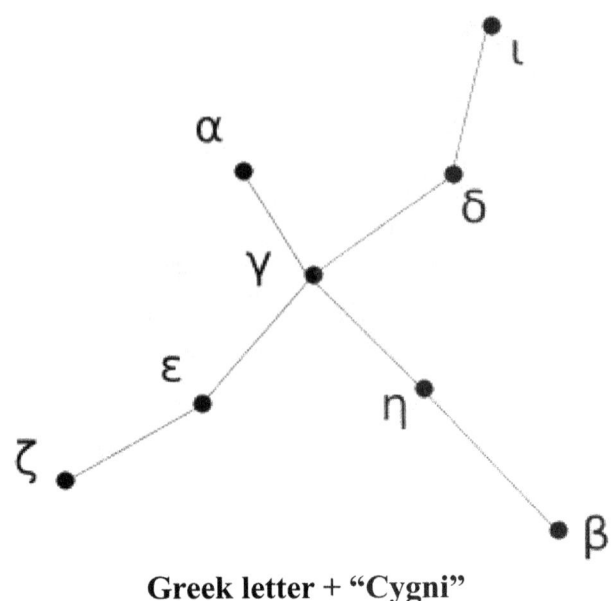

Greek letter + "Cygni"

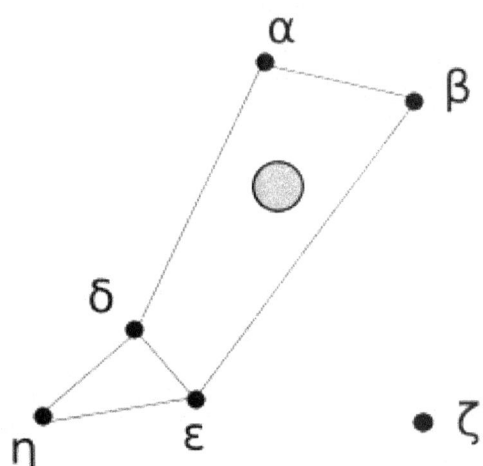

Greek letter + "Canis Majoris"

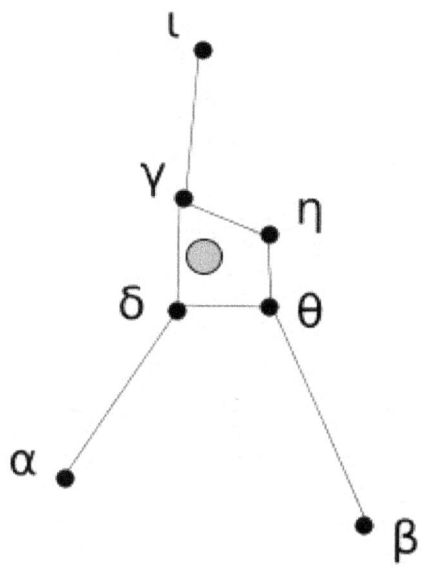

Greek letter + "Cancri"

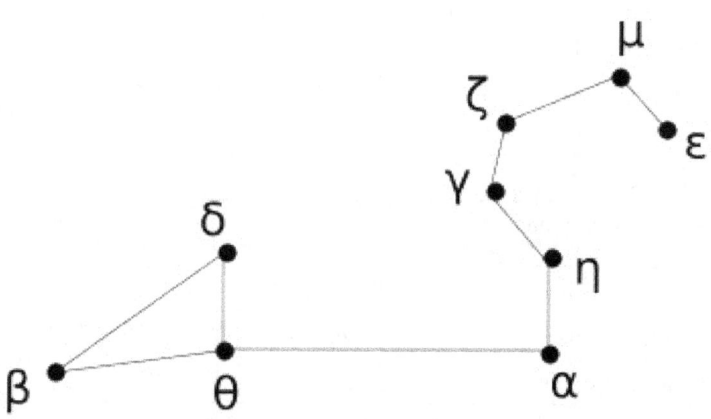

Greek letter + "Leonis"

Star Name	Common Name	Apparent Magnitude
Alpha Ursae Majoris	Dubhe	1.79
Beta Ursae Majoris	Merak	2.37
Gamma Ursae Majoris	Phecda	2.44
Delta Ursae Majoris	Megrez	3.31
Epsilon Ursae Majoris	Alioth	1.77
Zeta Ursae Majoris	Mizar	2.23
Eta Ursae Majoris	Alkaid	1.86
Alpha Ursae Minoris	Polaris	2.0
Beta Ursae Minoris	Kochab	2.1
Gamma Ursae Minoris	Pherkad	3.0
Delta Ursae Minoris		4.4
Epsilon Ursae Minoris		4.2
Zeta Ursae Minoris		4.3
Eta Ursae Minoris		5.0
Alpha Cassiopeiae	Schedar	2.5
Beta Cassiopeiae	Caiph	2.8
Delta Cassiopeiae	Ruchbah	2.8
Gamma Cassiopeiae		2.8
Epsilon Cassiopeiae		3.4
Alpha Leonis	Regulus	1.35
Beta Leonis	Denebola	2.11
Gamma Leonis	Algeiba	2.28
Delta Leonis	Zosma	2.56
Epsilon Leonis		3.0
Zeta Leonis	Adhafera	3.4
Eta Leonis		3.5
Theta Leonis	Chort	3.32

Star Name	Common Name	Apparent Magnitude
Alpha Orionis	Betelgeuse	0.5
Beta Orionis	Rigel	0.12
Gamma Orionis	Bellatrix	1.64
Delta Orionis	Mintaka	2.23
Epsilon Orionis	Alnilam	1.7
Zeta Orionis	Alnitak	2.05
Lambda Orionis	Meissa	3.54
Kappa Orionis	Saiph	2.06
Alpha Lyrae	Vega	0.03
Beta Lyrae	Sheliak	3.5
Gamma Lyrae	Sulafat	3.2
Delta Lyrae		4.3
Epsilon Lyrae		4.3
Zeta Lyrae		5.7
Alpha Cygni	Deneb	1.25
Beta Cygni	Albireo	3.08
Gamma Cygni	Sadr	2.23
Delta Cygni	Rukh	2.87
Epsilon Cygni	Gienah	2.48
Alpha Canis Majoris	Sirius	-1.46
Beta Canis Majoris	Mirzam	1.99
Delta Canis Majoris	Wezen	1.83
Epsilon Canis Majoris	Adhara	1.50
Zeta Canis Majoris	Furud	3.02
Eta Canis Majoris	Aludra	2.45

Stars can be classified according to their surface temperature.

Class	Surface Temp. (K)	Approximate Color
O	28,000 – 50,000	Blue
B	9,900 – 28,000	Blue / White
A	7,400 – 9,900	White
F	6,400 – 7,400	White / Yellow
G	4,900 – 6,000	Yellow
K	3,500 – 4,900	Orange
M	2,000 – 3,500	Red

There are a few stars every budding astronomer should know.

Star	Surface Temp. (K)	Class	Magnitude
Sun	5,800	G	-26.74
Sirius	9,940	B	-1.46
Betelgeuse	3,500	K	0.42
Rigel	11,000	B	0.12
Vega	9,600	A	0.03
Polaris	7,200	F	1.97

Stars appear to twinkle. If you do not believe me, go outside during the next clear night you have. Look at a bright star. Really look at it. You will notice that it twinkles. It does this because our atmosphere contains chemicals that mix and flow, similar to how water mixes and flows in a river or ocean. This mixing and flowing disturbs the light that leaves an object. Look at an object in the bottom of a river or pool and you will notice that image is unsteady. Similarly, stars appear to twinkle as their light passes through Earth's atmosphere.

I feel obligated to talk just a little bit about how stars die. The more massive a star is, the more interesting it is when it dies.

"Lighter" stars (stars with less mass), like our sun, eventually convert all of the hydrogen in their cores into helium. When this happens, gravity forces the core to collapse (the gravity was always there, but the pressure from the nuclear core was keeping the star from collapsing). As the star collapses, the hydrogen in the next outer layer begins to start undergoing nuclear fusion to form helium. These nuclear reactions happen much faster, so the resultant pressure is powerful enough to push the outer layers outward. This makes the star very big. These stars are called red giants and they will appear red (even to the naked eye). Arcturus and Aldebaran are two popular examples.

As the core continues to contract, the helium will eventually become so dense that it actually starts to create carbon and Oxygen nuclei through nuclear fusion. In these "lighter" stars, the process stops there, and when all the helium converts into carbon and oxygen, the core is called a white dwarf star (even though no nuclear reactions are happening). The gases in the outer layers begin to spread out into the universe and fluoresce because of the UV light emitted from the white dwarf core. This is called a planetary nebula. This will eventually happen to our sun. M27 is a great example.

"Heavier" stars are a little more dramatic. When we say "heavy" we mean approximately 8 times as massive as our sun. The process is the same, except the cores will become so dense that they can convert the oxygen and carbon into even heavier elements (up to iron). The stars at this stage are called red super giants (Betelgeuse is a popular example). These huge stars (radii can be as large as several AUs) eventually explode as supernova and form neutron stars (stars with a less massive core) or black holes (stars with a more massive core). Sorry, you aren't going to find any black holes with your telescope!

Before we start talking about telescopes, we need to understand two basic physics principles.

The Law of Refraction

This law explains how light "bends" as it passes through different mediums. Every medium has a "index of refraction" (n) associated with it. You can determine the index of refraction of a material if you can calculate the speed of light passing through that material.

$$n = \frac{speed\ of\ light\ in\ vacuum}{speed\ of\ light\ in\ material} = \frac{c}{v}$$

Consider two materials "1" and "2" with index of refraction values n1 and n2, respectively. As light passes through the 1-2 barrier, it will bend. The law of refraction tells us how.

$$n_2 sin\theta_2 = n_1 sin\theta_1$$

We can consider three examples: (1) n1 is equal to n2, (2) n1 is greater than n2, and (3) n1 is less than n2.

When n2 is equal to n1, theta 2 will be equal to theta 1. That much is intuitive. The light will not bend.

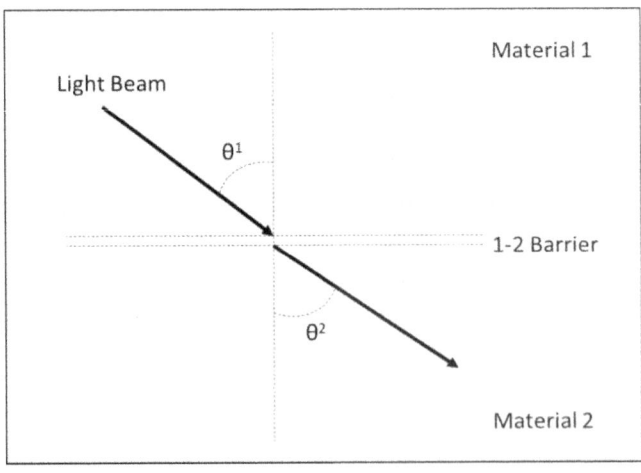

When n2 is greater than n1, theta 2 will be smaller than theta 1. The light will bend towards the normal in this scenario, so theta gets smaller.

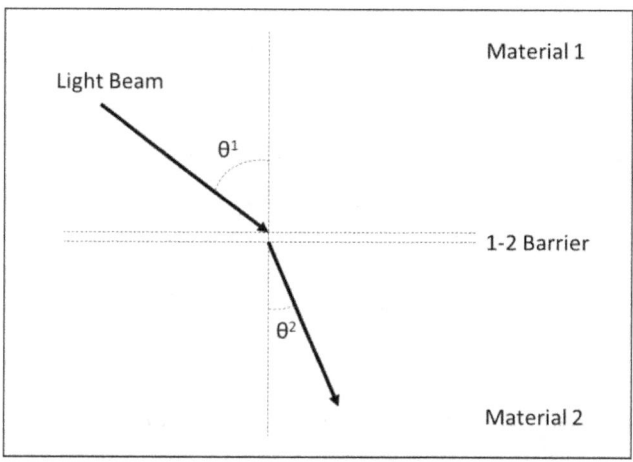

In the opposite example (n2 smaller than n1), theta 2 will be greater than theta 1 and the light will bend away from the normal.

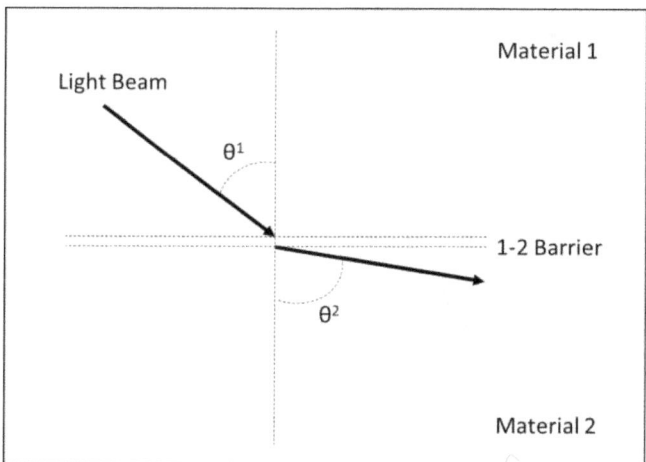

These telescope makers must be smart people! Imagine trying to design a refractor telescope's objective lens using the law of refraction!

The Law of Reflection

The law of refraction deals with mirrors. It simply states that theta 1 and theta 2 shown in the following image are always equal.

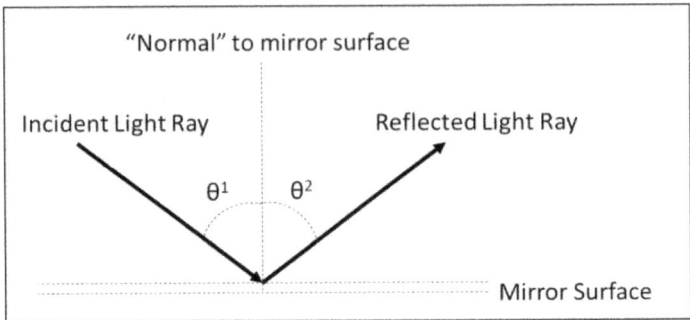

Things get very interesting when we consider curved mirrors, especially parabolic mirrors (curved mirrors with a parabolic shape).

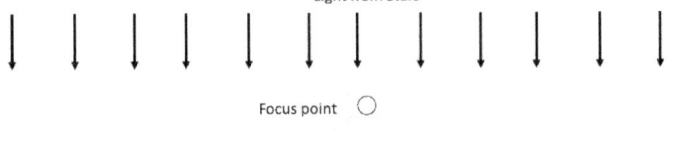

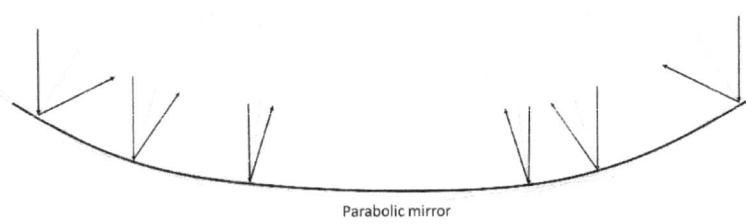

Each ray bounces off the curved mirror surface according to the law of reflection and all the rays meet at the focus point, where an image of stars is produced. All you have to do is put an eyepiece or camera sensor there to see them.

Part 2: Telescopes

The two most popular types of telescopes are: (1) the refractor and (2) the reflector. Reflectors have their many advantages, but they are slightly more complex than the simple refractor telescope. Refractor telescopes are usually simpler to use and cheaper to buy. This makes refractor telescopes perfect for the amateur astronomer. They are inexpensive and impressive scientific tools.

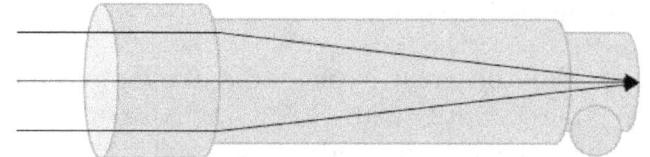

The Refractor Telescope

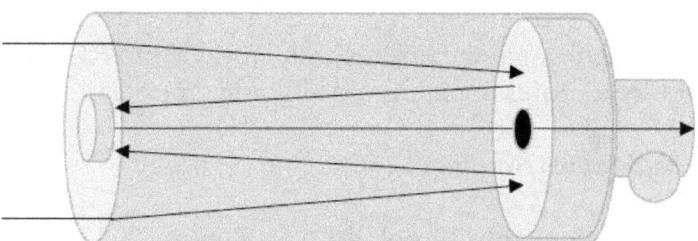

The Catadioptric Telescope

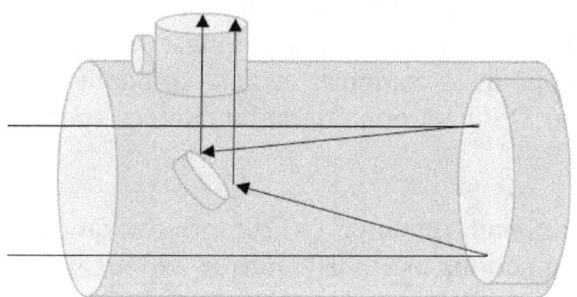

The Reflector Telescope

Refractor telescopes have an objective lens in the front that takes in the light from a distant object and focuses it to a bright point somewhere in the telescope tube. Refraction occurs in the objective lens.

Newtonian reflector telescopes have a primary mirror (mounted in a primary mirror cell at the base of the telescope) that directs the light from a distant object to a secondary mirror (mounted on something called a spider) that focuses the light to a bright point. Reflection occurs on the primary and secondary mirrors. The primary mirror in a reflector telescope should be parabolic and the secondary should be as flat as possible.

There are a few different types of Catadioptric telescopes. Personally, I am familiar with SCT-type, Maksutovs, and RC telescopes. Each have their own advantages and disadvantages. For the simple astronomer, Maksutovs are nice because they are portable and require little maintenance. In my opinion, the biggest advantage of these telescopes is that they provide longer focal lengths in a much smaller package. Stay away from RC telescopes unless you're a professional mirror alignment master.

For the budget astronomer...

DSO Astrophotography: Look for a 50-80mm f/5. Keep it cheap for now. Bigger is almost never better here.

Planetary/Lunar Astrophotography: You can start with a 5" Maksutov if you want something compact. An 8" SCT is another good starting point. A computerized alt-az mount is a good idea if you never do DSO astrophotography. Bigger aperture is usually better here.

Visual use: Small refractor (<120mm) for quick sessions. 8" dobsonian for serious use. A 10" mirror will resolve more stars in globular clusters if they are something you like looking at. Bigger aperture is always better here but a 12" telescope is huge.

Chapter 4

The Reality of the Cheap Telescope

I want to take a quick reality check before we move on.

For the sake of conversation, I am going to assume you purchased or own a cheap, "small" telescope. What makes a telescope "small" anyway? When astronomers refer to "small" telescopes, they are usually referring to telescopes with an aperture of 100 millimeters (~4 inches) or less. Just because you bought a cheap, small telescope does not mean you necessarily made a mistake. The telescope you are working with is most likely *much* better than the telescopes astronomers were using in the seventeenth and eighteenth centuries (but you must deal with light-polluted skies). In fact, not only should you *not* be discouraged by your cheap telescope – you should be excited about it!

Look up some images that the Hubble space telescope has taken. Do not expect anything remotely close to those images with a cheap telescope (or any telescope on the surface of the Earth). Some deep sky objects may even be impossible to see with a small telescope in a light-polluted area. Light pollution is a term that refers to the light that comes from nearby civilization that interferes with astronomical observations. If you live in a light-polluted area, you will have a hard time seeing deep sky objects like nebula and galaxies (depending on the quality and size of your telescope).

That being said, devoting time to mastering a small telescope is definitely worth it. You can still observe bright nebula, some planets, and the moon with an aperture smaller than 3 inches (~80mm). Galaxies, nebula, and globular clusters will most likely look like faint, fuzzy glows of light. It may be difficult to resolve individual stars in globular clusters with your small aperture.

You are most likely going to get frustrated with visual astronomy and end up pursing astrophotography. We will discuss that later.

M42 as seen visually ^

M42 using affordable astrophotography equipment ^

Chapter 5

Basic Telescope Science & Mathematics

A refractor telescope has one large lens in the front (the objective lens) and an eyepiece that magnifies the image so you can see it. The size (specifically the diameter) of the objective lens (or primary mirror) is called the **aperture**. The aperture determines how much light your telescope can "take in". The larger the aperture, the better (see chapter 11).

I own and use a small 50mm refractor telescope. The aperture is given in the name – the diameter of the objective lens is 50mm (50 millimeters). The focal length of the objective lens is 600 millimeters. When you go to buy a telescope, the aperture and the focal length of the objective lens are two things you should definitely consider.

The **resolution**, *R*, of a telescope determines how much detail you can make out in the image. The resolution can be estimated using the following equation. This equation returns a value in arc-seconds (the value of the aperture must be entered in millimeters).

$$R = \frac{120}{Aperature}$$

A 50mm refractor has a resolution of ~2.4 arc-seconds. There is nothing you can do to improve that. This is one reason why the aperture of a telescope is so important. A resolution of 2.4 arc-seconds is generally not adequate for detailed planetary observations.

Table 5.1 (next page) lists the resolutions of the 50mm, 60mm, 70mm, 80mm, 90mm, and 102mm refractor telescopes. Please note that this equation is also valid for reflector telescopes.

Telescope	Resolution (arc-seconds, degrees)
50mm refractor	2.4, 0.000667
60mm refractor	2, 0.000556
70mm refractor	1.71, 0.000475
80mm refractor	1.5, 0.000417
90mm refractor	1.33, 0.000369
102mm refractor	1.18, 0.000328

Table 5.1 – The resolutions of typical refractor telescopes.

The **maximum useful magnification** a telescope can handle is (somewhat) directly related to the aperture of the telescope you are using. For a *small refractor*, the maximum useful magnification is commonly given as 2.5 times magnification per millimeter of aperture (or the aperture in millimeters multiplied by 2.5). If you try to use a magnification higher than this value, you may not be able to focus the image. This is related to the resolution of a telescope. When you use a magnification higher than what your telescope can handle, you are magnifying more than what your telescope can resolve. You cannot focus the image because your telescope cannot resolve the image.

According to this math, a 50 mm refractor would have a maximum magnification of 125 times; a 70mm refractor would have a maximum magnification of 175 times; a 80mm refractor would have a maximum magnification of 200 times; a 90mm refractor would have a maximum magnification of 225 times; and so on.

In practice, however, this is a slightly outdated equation. Refractor telescopes with a small aperture are not restricted to this limit, while other types of telescopes may not be able to handle the maximum magnification predicted by this equation (the maximum useful magnifications of reflector telescopes is commonly given as 30-40 times per inch of aperture).

In 2017, the maximum useful magnification a small refractor telescope can achieve depends more on the quality of its construction and the steadiness of the atmosphere when it is being used. I have heard stories of quality 80mm telescopes achieving as high as 250 times magnification on extremely steady nights.

The more you magnify, the more dim the object you are observing becomes. Because of this, you do not always want to use high magnifications. During average seeing conditions, you may find it best to use magnifications closer to 30 times per inch of aperture for planets; 50 times per inch of aperture for stars; and 15 times per inch of aperture for clusters and nebulae. Obviously, these are very rough guidelines. Experiment enough and you will figure out what works best for you and your setup.

You will be able to use higher magnifications when the skies are steady. Steady nights are rare, so prepare to wait.

The **magnification** you are observing with at any time depends on the focal length of the objective lens (or primary mirror) and the focal length of the eyepiece you are using. The focal length of the eyepiece is given in the name of the eyepiece. For example, a 15mm eyepiece has a focal length of 15mm. The magnification can be calculated using the following equation.

$$Magnification = \frac{f_{OBJECTIVE}}{f_{EYEPIECE}} = \frac{f_{OBJ}}{f_{EP}}$$

Where f is the focal length. My 50mm telescope came with three eyepieces: a 20mm, a 10mm, and a 4mm. The 20mm eyepiece gives 30 times magnification (600/20), the 10mm eyepiece gives 60 times magnification (600/10), and the 4mm eyepiece gives 150 times magnification (600/4).

According to the previous maximum useful magnification equation, my 50mm refractor can only handle magnifications up to 125 times. I have actually been able to use the 4mm eyepiece that came with the telescope on steady nights. To figure out how much magnification your telescope can actually handle, you have to employ the classic trial-and-error method.

I can use the 4mm eyepiece that came with my 50mm telescope despite the fact it gives a magnification higher than what the telescope is expected to be able to handle. I have tried more powerful eyepieces and I cannot get the image to focus. I can, therefore, state that the maximum magnification my 50mm telescope can handle is approximately 150 times during a very steady night sky.

All you can do is test your eyepieces and see which ones work. If you cannot get the image to focus: (1) the eyepiece is too powerful for that telescope, (2) the sky is not steady enough for such a powerful magnification, or (3) the eyepiece is a cheap, worthless eyepiece.

It should be noted that the unsteadiness in the sky limits magnification to around 300-400 times. I do not expect that will be a problem for us. In fact, any magnification over 200 times is really not necessary. When you use higher magnifications, you typically sacrifice contrast and detail.

I accept the fact that many of you may hate math. Tables 5.2, 5.3, 5.4, 5.5, 5.6, 5.7, 5.8, and 5.9 list the magnifications of certain telescope-eyepiece combinations.

Magnification Tables
(Found on the following pages)

Eyepiece (mm)	Magnification (x)
40	10
36	11
30	13
25	16
20	20
19	21
17	24
15	27
14	29
12	33
11	36
10	40
9	44
8	50
7	57
6	67
5	80
4	100
3	133
2	200
1	400

Table 5.2 – Magnification of various telescope-eyepiece combinations. F_{OBJ} = 400mm.

Eyepiece (mm)	Magnification (x)
40	13
36	14
30	17
25	20
20	25
19	26
17	29
15	33
14	36
12	42
11	45
10	50
9	56
8	63
7	71
6	83
5	100
4	125
3	167
2	250
1	500

Table 5.3 – Magnification of various telescope-eyepiece combinations. $F_{OBJ} = 500$mm.

Eyepiece (mm)	Magnification (x)
40	15
36	17
30	20
25	24
20	30
19	32
17	35
15	40
14	43
12	50
11	55
10	60
9	67
8	75
7	86
6	100
5	120
4	150
3	200
2	300
1	600

Table 5.4 – Magnification of various telescope-eyepiece combinations. $F_{OBJ} = 600$mm.

Eyepiece (mm)	Magnification (x)
40	18
36	19
30	23
25	28
20	35
19	37
17	41
15	47
14	50
12	58
11	64
10	70
9	78
8	88
7	100
6	117
5	140
4	175
3	233
2	350
1	700

Table 5.5 – Magnification of various telescope-eyepiece combinations. F_{OBJ} = 700mm.

Eyepiece (mm)	Magnification (x)
40	20
36	22
30	27
25	32
20	40
19	42
17	47
15	53
14	57
12	67
11	73
10	80
9	89
8	100
7	114
6	133
5	160
4	200
3	267
2	400
1	800

Table 5.6 – Magnification of various telescope-eyepiece combinations. $F_{OBJ} = 800mm$.

Eyepiece (mm)	Magnification (x)
40	23
36	25
30	30
25	36
20	45
19	47
17	53
15	60
14	64
12	75
11	82
10	90
9	100
8	113
7	129
6	150
5	180
4	225
3	300
2	450
1	900

Table 5.7 – Magnification of various telescope-eyepiece combinations. F_{OBJ} = 900mm.

Eyepiece (mm)	Magnification (x)
40	25
36	28
30	33
25	40
20	50
19	53
17	59
15	67
14	71
12	83
11	91
10	100
9	111
8	125
7	143
6	167
5	200
4	250
3	333
2	500
1	1000

Table 5.8 – Magnification of various telescope-eyepiece combinations. $F_{OBJ} = 1000mm$.

Eyepiece (mm)	Magnification (x)
40	28
36	31
30	37
25	44
20	55
19	58
17	65
15	73
14	79
12	92
11	100
10	110
9	122
8	138
7	157
6	183
5	220
4	275
3	367
2	550
1	1100

Table 5.9 – Magnification of various telescope-eyepiece combinations. $F_{OBJ} = 1100$mm.

Eyepiece (mm)	Magnification (x)
40	30
36	33
30	40
25	48
20	60
19	63
17	71
15	80
14	86
12	100
11	109
10	120
9	133
8	150
7	171
6	200
5	240
4	300
3	400
2	600
1	1200

Table 5.10 – Magnification of various telescope-eyepiece combinations. $F_{OBJ} = 1200mm$.

Eyepiece (mm)	Magnification (x)
40	33
36	36
30	43
25	52
20	65
19	68
17	76
15	87
14	93
12	108
11	118
10	130
9	144
8	163
7	186
6	217
5	260
4	325
3	433
2	650
1	1300

Table 5.11 – Magnification of various telescope-eyepiece combinations. $F_{OBJ} = 1300$mm.

Eyepiece (mm)	Magnification (x)
40	35
36	39
30	47
25	56
20	70
19	74
17	82
15	93
14	100
12	117
11	127
10	140
9	156
8	175
7	200
6	233
5	280
4	350
3	467
2	700
1	1400

Table 5.12 – Magnification of various telescope-eyepiece combinations. $F_{OBJ} = 1400mm$.

Eyepiece (mm)	Magnification (x)
40	38
36	42
30	50
25	60
20	75
19	79
17	88
15	100
14	107
12	125
11	136
10	150
9	167
8	188
7	214
6	250
5	300
4	375
3	500
2	750
1	1500

Table 5.13 – Magnification of various telescope-eyepiece combinations. $F_{OBJ} = 1500mm$.

Why don't we go ahead and plot some of this data.

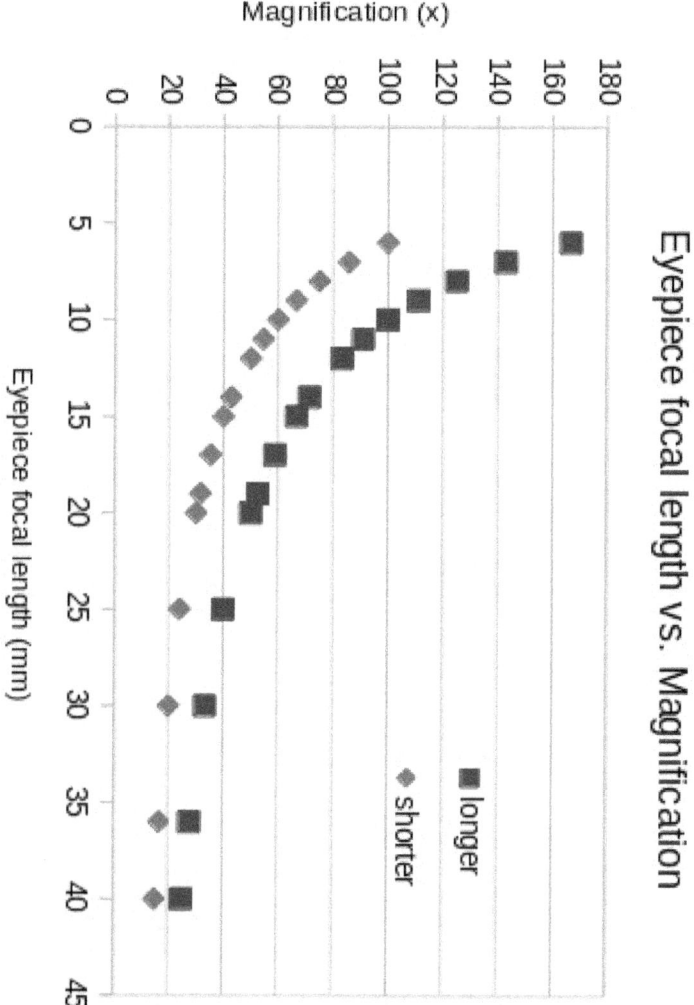

We can see that the magnification seems to increase exponentially as the eyepiece's focal length approaches smaller values (mathematically speaking, there is an asymptote where the focal length of the eyepiece equals 0). The data labeled "longer" refers to a telescope with an objective lens with a longer focal length.

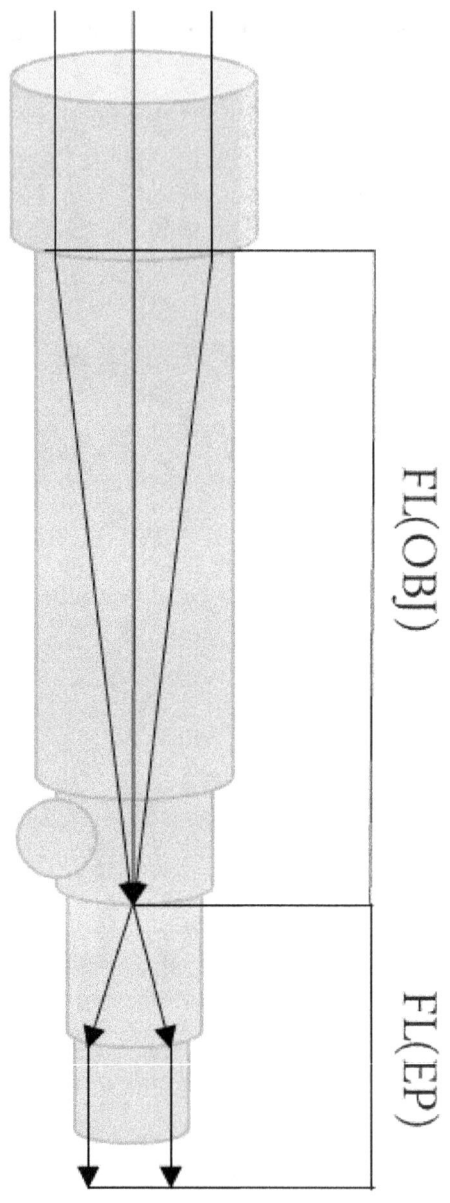

Figure 5.1 – The focal lengths of the objective lens (OBJ) and the eyepiece (EP)

A discussion on the **light gathering ability** of a telescope can be found in chapter 11. It is included there because it greatly dictates what telescope you should buy.

When you go to buy a telescope, you will notice that the "**f number**" is usually listed (in the form *f/x,* where *x* is the f number). It is simply equal to the ratio of the focal length of the objective lens to the diameter of the objective lens. Astronomers typically like to use millimeters as the unit of length in these equations. You can use whatever units you want, but just make sure both values are expressed in the same units. In other words, the f-number is a dimensionless value.

$$f\# = \frac{(f_{OBJ})}{aperature}$$

For example, my 50mm refractor is a f/12 telescope since the f number is equal to 600/50, or 12. When you go to buy a telescope, they always list the aperture of the telescope in the description. They also list either the f number, the focal length of the objective lens, or both. Obviously, they do not need to list both since they can be easily calculated from one another if the aperture is known (which it always is).

If the f number is roughly in the 4 to 5 range, it is best for deep sky observations; if it is over 9 or 10, it is best for lunar and planetary observations; if it is somewhere in the 5 to 9 range, it works well with both. That being said, every telescope can observe nebula and planets. The f number just determines the quality of those observations. Sometimes, a telescope with a small f number (under 5) is called "fast". If the f number is larger than ~15, color dispersion will be greatly reduced. This is ideal for lunar, planetary, and double star observations.

Some people argue that the f number is really only a relevant term in astrophotography, but this is simply false.

Smaller f numbers are better for deep sky object observations because shorter telescopes magnify less and show more space (a larger field of view) when compared to longer telescopes using the same eyepiece.

Larger f numbers are better for lunar and planetary observations because they magnify more and eliminate noticeable **chromatic aberration**. Optics can refract certain wavelengths of light more so than others, and this can cause different colors of light to focus at different points. This explains why a shorter telescope will make bright objects appear to have a purple or yellow edge (like the limb of the moon). In the following image, blue light is represented by the hashed line and red light is represented by the straight line.

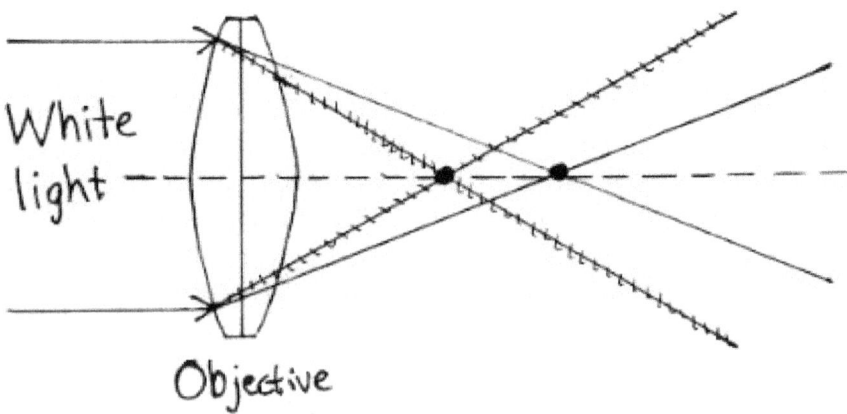

Issac Newton developed the reflector telescope to eliminate this effect (which he assumed to be present). So what? Are all refractor telescopes worthless? Absolutely not. Refractor telescopes can be improved in a variety of ways. The simplest way is to make the telescope longer (use an objective lens with a longer focal length). This decreases the distance between the points different wavelengths of light focus. This is one reason why larger f numbers (larger than 12) are recommended for lunar and planetary observations – the telescopes are longer.

There are two popular styles of refractor telescopes: (1) the achromatic refractor and (2) the apochromatic (APO) refractor. Both lessen the effect of chromatic aberration.

A popular achromatic refractor, called an achromatic doublet, brings the focus points of red and blue light to the same location. This is accomplished by using two different types of glass (typically crown and flint) that refract blue and red light differently. This corrects chromatic aberration, but it does not eliminate it. Blue light is represented by the hashed line and red light is represented by the straight line in the following image.

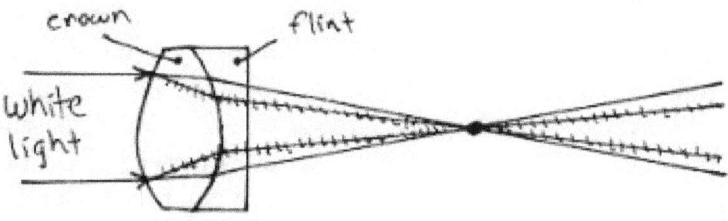

An APO refractor uses more than two lenses in the objective to bring the focus points of more than two different wavelengths of light (for example: red, blue, and green) to the same location. APO refractor telescopes correct chromatic aberration better than achromatic telescopes. Blue light is represented by the hashed line in the following image.

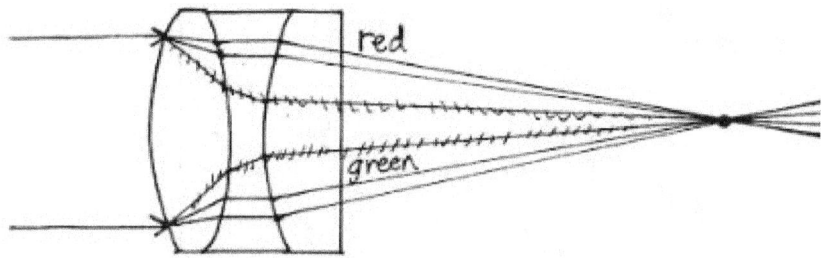

Most refractor telescopes are achromatic. APO refractor telescopes are fairly expensive (since they are better).

I stated earlier that "Some people argue that the f number is really only a relevant term in astrophotography", why is that? The f number tells us the relative speed of the optical system. A F/5 is faster than a F/6. How much does this matter? It has some implications in planetary/lunar imaging, which we will get to, but it matters much more in DSO imaging.

This speed is a big deal in DSO imaging. It tells us how quickly we gather photons to produce a bright image. A faster telescope will require lower exposure times, which is good. It lets you take 16, 15-minute photos (4 hours total) instead of 16, 1-hour photos (16 hours total) if 16 sub-exposures meet your stacking goals. You can get 16, 15-minute sub-exposures done in one night, even if planes or satellites mess up a few. If a plane messes up even one of your 1-hour sub-exposures, you just wasted a whole hour as opposed to 15 minutes with the faster system. These things add up quickly. Let's do some math. The required exposure time (Exp) equivalent to a faster optical system can be calculated via a ratio calculation.

$$Exp_{Slow} = \frac{F_{Slow}^2}{F_{Fast}^2} * Exp_{Fast}$$

F/X	Ratio	Equivalent Exposure Time (F/4 Basis) (Minutes)
2	0.25	2.50
3	0.56	5.63
4	1	10.00
4.5	1.27	12.66
5	1.56	15.63
5.5	1.89	18.91
6	2.25	22.50
6.5	2.64	26.41
7	3.06	30.63
7.5	3.52	35.16
8	4	40.00
9	5.06	50.63
10	6.25	62.50

Arc-seconds, Arc-minutes, and Degrees

Do not let these terms intimidate you. These terms are used to define how much detail a telescope can reveal (resolution), how large an object is in the sky, and the separation between two objects, particularly stars in multiple star systems.

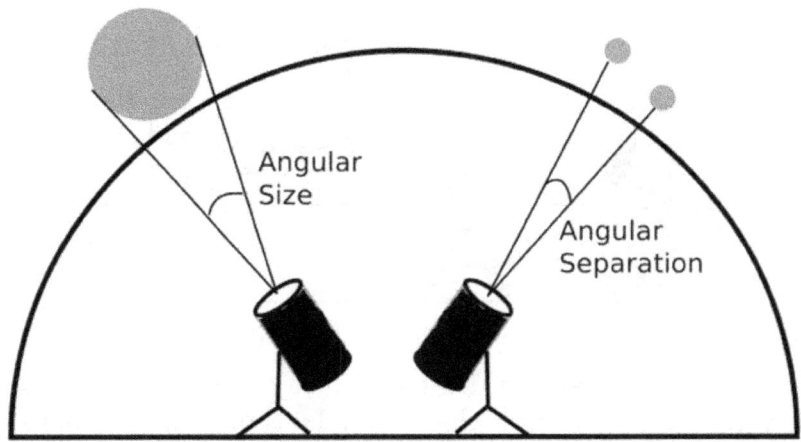

Values in degrees are written as X°; values in arc-minutes are written as X'; and values in arc-seconds are written as X", where X is the value. Some of you may already know that 1 degree (1°) is equal to 1/360 of a circle. What is less commonly known is that 1 arc-minute (1') is equal to 1/60 of a degree and that 1 arc-second (1") is equal to 1/3600 of a degree. Notice how the relationship between degrees, arc-minutes, and arc-seconds is similar to the relationship between hours, minutes, and seconds.

1 degree = 60 arc-minutes = 3600 arc-seconds
1 hour = 60 minutes = 3600 seconds

Your fist at an arm's length away takes up approximately 10 degrees of the night sky. Ursa Major takes up approximately 20 degrees of the night sky. Go outside and see if your fist, at an arm's length away, takes up half of Ursa Major.

In the case you're still a little confused, let's continue to think about these terms. If you're not, go ahead and skip to the next chapter. Imagine standing in a room with a television and a vase on a coffee table.

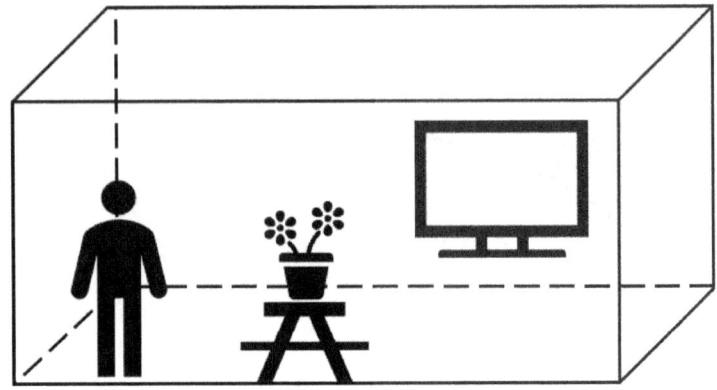

Hold out your first at an arm's length away. There is a real chance that your fist can entirely block the television, but only partially block the vase. This means that, from your perspective, the apparent size of the vase is greater than the apparent size of the television! Obviously, the television is bigger than the vase, but from your perspective, the apparent size of the vase is greater.

Arc-seconds, arc-minutes, and degrees help us quantify these apparent sizes. At an arm's-length away:
- Your fist takes up ~10 degrees
- Your middle three fingers take up ~5 degrees
- Your pinky finger takes up ~1 degree (or 60arc-minutes)

In this example, your television could be 10 degrees in size, but that vase could nearly 14 degrees in size (1 fist + 2 thumbs).

These angular sizes give a quantitative measurement for how big something appears to us from our location. This is, obviously, very important for astronomy, but possibly not so important for your living room.

Now imagine that you are in boundless, expanding universe (which you are, whether you like it or not). This universe contains yourself, the moon, and Saturn.

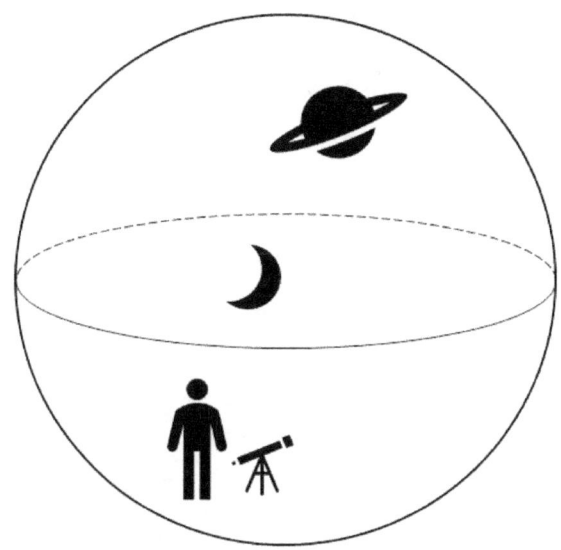

We already know that the diameter of the moon is only 2,000 miles compared to Saturn's diameter of 72,000 miles, yet the moon looks much bigger than Saturn. Here, Saturn is the television and the moon is the vase.

The moon looks bigger since it is closer to us. It has a larger angular size, despite having a smaller actual size compared to Saturn.

	Diameter (Miles)	Angular Size (degrees)	Angular Size (')	Angular Size (")
Sun	864,576	0.53	31.8	1908
Moon	2,159	0.52	31.2	1872
Mars	4,212	0.00695	0.417	25
Jupiter	86,881	0.01385	0.831	50
Saturn	72,367	0.00574	0.3444	21

Chapter 6

Parts and Accessories

The Eyepiece

The eyepieces that come with cheap telescopes are not usually high-quality eyepieces. Quality eyepieces are not *too* expensive, so it is worth looking into purchasing better ones (it makes a huge difference). There is a lot that goes into choosing eyepieces, but it is probably best to go with eyepieces that get good reviews and are reasonably priced. Just make sure you do not buy an eyepiece that will give a magnification greater than what your telescope can handle (see figure 6.3).

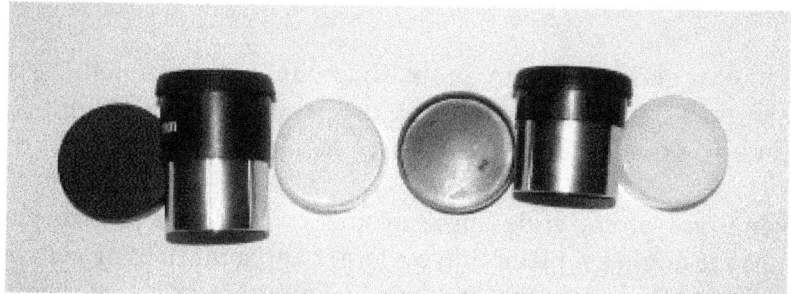

Figure 6.1 – 15mm (left) and 9mm (right) eyepieces with their protective caps.

When you are not using your eyepieces, make sure the protective caps are on them. You do not want dust, dirt, or moisture to get onto the optics. Cleaning optics is no easy task. This is similar to a camera lens. Always put the cap on when not in use.

If your telescope is large enough to handle them, 2" eyepieces are wonderful and allow for huge fields of view. It is always nice to look through a big piece of glass!

If you wear glasses, you should be concerned about the eye relief of the eyepiece.

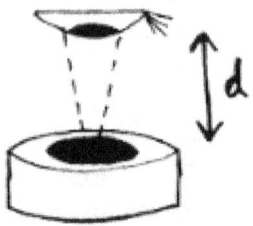

The eye relief determines how far away your eye has to be from the eyepiece to see the image. If you wear glasses, you will need a decent eye relief so that you can look into your telescope with your glasses on. In the image, *d* will be larger when the eye relief is larger.

If you wear glasses, you are probably going to want something with greater than 17mm of eye relief. Most people feel that 20mm is the minimum. I don't wear glasses, but I would go for 20mm if I did.

Personally, I prefer wide angle eyepieces (and you probably will too). Wide angle eyepieces have a larger apparent field of view, and this essentially means that you can see more space when you look into the eyepiece. Compare eyepieces *a* and *b* in the following image.

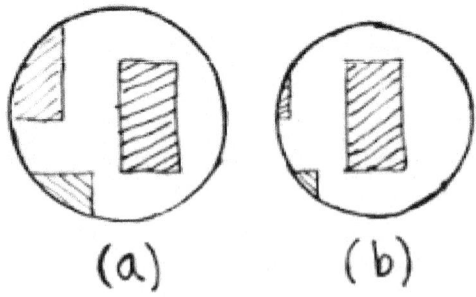

Figure 6.2 – Two different field of views.

Eyepiece *a* has a larger apparent field of view. Both eyepieces give the same magnification (assume they are both 20mm eyepieces), but *a* lets you see more. Notice how both eyepieces allow you to see the full rectangle on the right, but only eyepiece *a* lets you see more of the other shapes. Wide angle eyepieces are much more pleasant to use. This is mostly because they have a larger lens. Most cheap eyepieces have a very small hole you have to look through. A wide angle eyepiece makes using a telescope much easier and much more enjoyable. Typical wide angle eyepieces will contain 55°, 60°, 66°, or 72° in the description. This is simply the apparent field of view of the eyepiece. Personally, I like 62, but I have been tempted to get into the 80's for a few years now. *UPDATE* I have purchased a few 82's. They are neat, but 62 is fine. Eye relief is more important to me at this point. 20mm eye relief makes a very comfortable eye piece. Unfortunately, that is an expensive trait.

A longer discussion on field of view can be found in chapter 11 (we need to discuss deep sky objects first).

All accessories have an associated barrel diameter and this determines if a part will work with the telescope you have. If an eyepiece has a barrel diameter of 1.25" then it will not work with a telescope that requires accessories with a barrel diameter of 2" (unless you have an adapter). Most small telescopes require 1.25" barrel diameters (avoid any telescope that requires smaller barrels). If you own a telescope that requires 1.25" barrel diameter, make sure any eyepieces, filters, Barlow lenses, and any other accessories you may buy have that associated barrel diameter.

For planetary viewing, cheaper eyepieces may be better. The less glass, the better. This usually means cheaper and this is good for us. Go for an AFOV of 52-62 degrees with an eye relief of >17mm ideally. 200x magnification is good for most nights. 250 is OK for Saturn, Mars, and the moon.

For DSOs, go crazy with AFOV, but make sure eye relief is still good (aim for 20mm here).

The Barlow Lens

A Barlow lens is a part that multiplies the magnification power of an eyepiece. If your cheap telescope came with a Barlow lens, it may or may not work (I have not had good experiences with the Barlow lenses that came with my telescopes). A good Barlow lens is a nice tool to have because you will not have to purchase as many eyepieces (see figure 6.3). A 2x Barlow lens makes a 26mm eyepiece a 13mm eyepiece; a 9mm eyepiece a 4.5mm eyepiece; a 6mm eyepiece a 3mm eyepiece, and so on. A Barlow lens allows you to look through the bigger lenses associated with less powerful eyepieces with the magnification power of more powerful eyepieces.

I use a 2x Barlow lens. This doubles the magnification power of any eyepiece I have. There are other Barlow lenses that multiply the power by 3 times, 4 times, 5 times, and so on. A quality Barlow lens is still relatively cheap. Some people warn amateurs about Barlow lenses that multiply power by more than 2 times.

If you see something with good reviews and it costs more than $40.00, I think it would serve you well.

The Diagonal

A diagonal allows you to view the image at a 90 degree angle to the direction the telescope is pointing. This makes viewing objects easier and more comfortable. One should come with your cheap telescope (see figure 6.3). Refractor, SC, and Maksutov telescopes use diagonals. Get a dielectric one if you want to invest in a lifetime diagonal.

When you look into a telescope, the image may be inverted in some fashion. There are types of diagonals (star diagonals) and prisms that correct this effect, but they are not necessary (in my opinion). Once you get used to the effect, you will not worry about it. Figure 6.3 shows an image I took using a telescope and Figure 6.4 shows the "corrected" version.

Figure 6.3 – Raw image from telescope.

Figure 6.4 - "Corrected" image.

Putting it all together

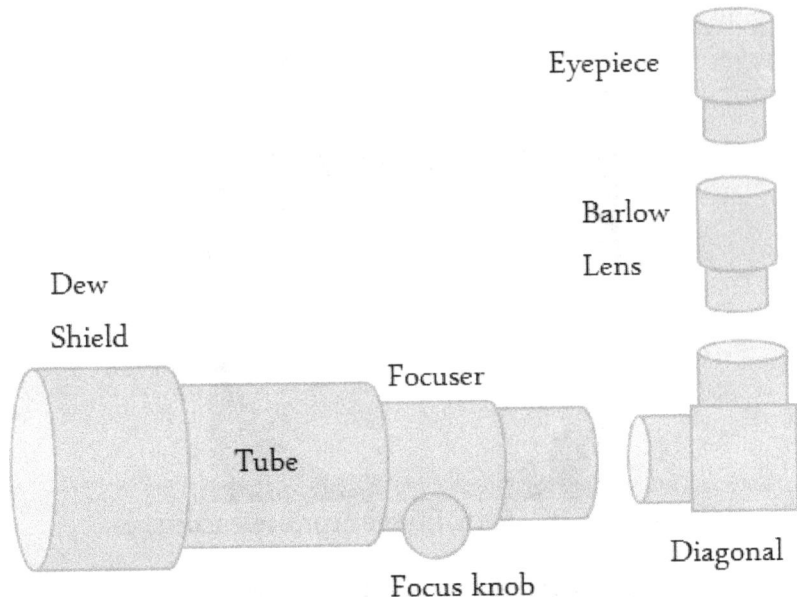

Figure 6.5 – The parts of a refractor telescope

You will have to spin the knob to focus the image throughout your observation sessions to focus the image. Spinning the knob causes the focuser drawtube to increase and decrease in length, depending on the direction of spinning.

Figure 6.6 describes the typical installation procedure for all these parts. The diagonal typically gets inserted onto the shaft that moves when you spin the focus knob. A Barlow lens (if one is used) gets inserted into the diagonal. An eyepiece gets inserted into the Barlow lens if a Barlow lens is used. If a Barlow lens is not used, then the eyepiece gets inserted into the diagonal.

Usually thumb screws are used throughout the assembly. First, you should loosen the thumb screw(s) all the way so that the piece of equipment (eyepiece, Barlow lens, etc.) you are inserting can be inserted easily (if you have to force it, something is wrong). After the piece of equipment is inserted, tighten the thumb screw so that the installation is secured and the equipment will not fall out. Do not over-tighten the thumb screws on a cheap telescope. Cheap telescope components are made of cheap plastic (so that the telescope can be sold for a low price). Over-tightening thumb screws can strip the plastic threads the thumb screws thread through. Tighten just so that it feels secure.

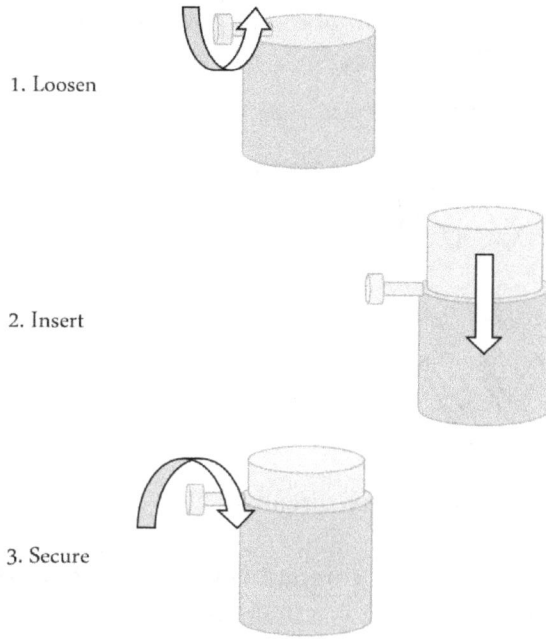

1. Loosen

2. Insert

3. Secure

Figure 6.6 – Inserting a telescope accessory using thumb screws. Do not over-tighten, but tighten enough so that the accessories will not fall out.

For imaging setups, you should try to use all threaded components. Threaded components remove some of the possibility for tilt induced by using thumbscrews. For example, you want to thread a field flattener into the focuser drawtube if possible.

Astrophotography Threads

A thread needs a diameter and a pitch to be defined. For example, a ¼-20 thread has a ¼" diameter and 20 threads per inch (1/20" pitch). A thread can either be female or male. You screw a male thread into a female thread as shown in the following image.

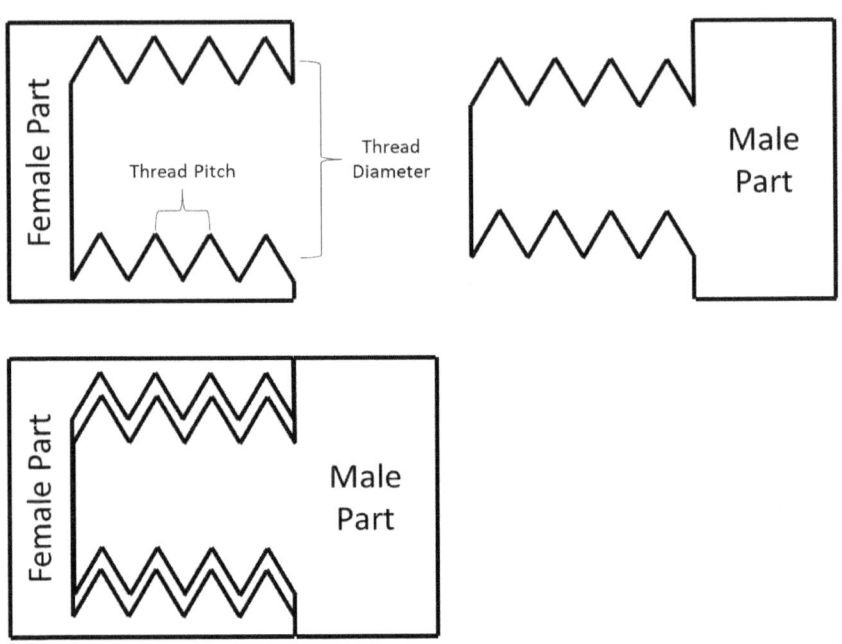

Common astrophotography threads:

Description	Thread Diameter	Thread Pitch
1.25" Filters	28.5mm	0.6mm
2" Filters	48mm	0.75mm
T2 Thread	42mm	0.75mm
"Wide" T Thread	48mm	0.75mm
Maksutov Back	44.5mm	1mm
Small SCT Back	2"	24 TPI
Large SCT Back	3.25" or 3.28"	16 TPI

Refractor drawtubes have various threads depending on the size and manufacturer. For astrophotography setups, you want to use all-threads and no thumbscrews if possible.

For example, you would thread the flattener into the focuser drawtube, thread any accessories into the flattener, and thread the camera into the accessories.

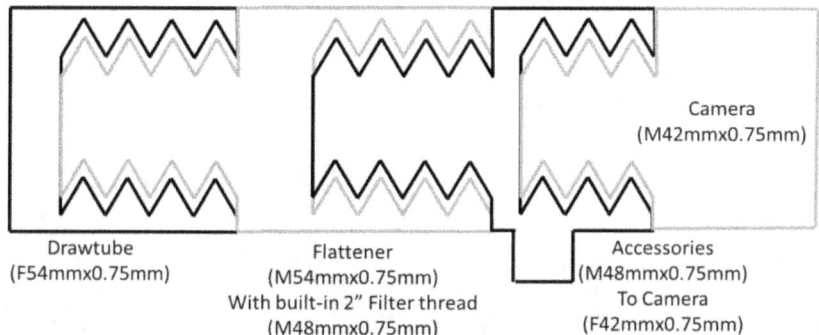

Drawtube
(F54mmx0.75mm)

Flattener
(M54mmx0.75mm)
With built-in 2" Filter thread
(M48mmx0.75mm)

Accessories
(M48mmx0.75mm)
To Camera
(F42mmx0.75mm)

Camera
(M42mmx0.75mm)

This "all-threaded" style is preferred for astrophotography. It may take some time for you to figure out the proper parts you need for your particular setup. For example, if you have a camera with male 42x0.75 threads, you need accessories that end with a female 42x0.75 thread.

Backspacing Spacers

We will talk about backspacing later, but you will likely need to purchase M42, M48, or even M54 spacers depending on your camera and particular astrophotography setup. These spacers have both male and female threads of the same type and only exist to yield the proper backspacing. For example, if your telescope flattener requires 55mm backspacing from the threads, then you need to figure out how to get that value. If you have 40mm built into your M42x0.75 camera and accessories, then you need a 15mm M42x0.75 spacer.

Here, you can see a pair of digital calipers used to measure the distance between the front of the camera and a reference point on the flattener. The drawings for the flattener and the camera allow for a calculation that yields a necessary measurement of 88.5mm.

The Finderscope

As far as finding objects goes, your telescope probably came with some type of finderscope. A finderscope is essentially a low power telescope. Theoretically, you can find the object you are hunting for with the finderscope and then proceed to look through the more powerful telescope. A finderscope must be correctly aligned. I find it easiest to do this during the day. Go outside and use your telescope to find an object, like a transformer on a distant telephone pole. Try to center the object using a high magnification. Then look through your finderscope and adjust however it is mounted to the telescope until the finderscope is also centered on that object. Only when your finderscope is correctly aligned will it be of any use. You may have to perform this "calibration" more often than you would like to.

For objects like the moon, it is easy to point and follow the bright light until you see it. For smaller objects like deep sky objects and planets, a finderscope is very useful if not necessary.

Your telescope may have also came with a red dot finder, but they work the same way. My 90mm refractor and my 130mm reflector came with a red dot finder. I actually prefer the red dot finder, but there are many people who do not. Since red dot finders do not magnify at all, it is more difficult to exactly point to the object you want to observe. A finderscope magnifies to some extent, so it is easier to point the telescope more accurately. What you prefer, however, is entirely up to you.

No matter what you use, finding objects is going to be frustrating. Stay patient, and you will find what you are looking for. It may be best to start with lower magnifications, find the object, and then switch to higher magnifications. After you get used to the frustration, hunting for deep sky objects becomes quite enjoyable. Equatorial mounts are confusing at first (since you cannot move up and down and side to side) but you will get used to it.

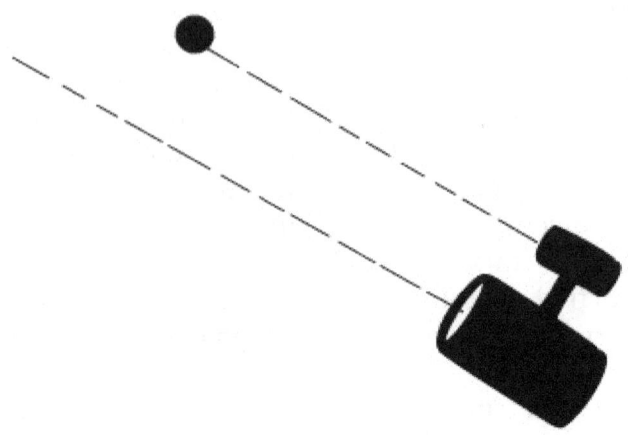

Figure 6.7 – An incorrectly aligned finderscope will not help you find objects with your telescope.

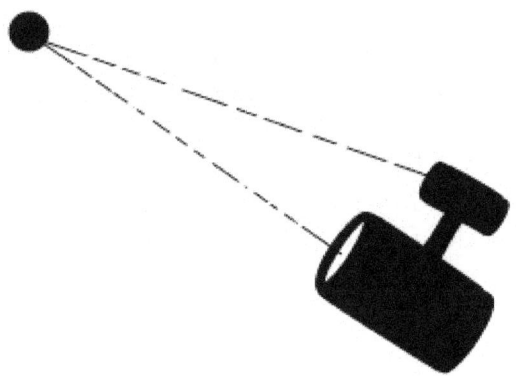

Figure 6.8 – A correctly aligned finderscope is very helpful.

You may find that it is best to use both a red dot finder and a finderscope. This combination gives you the ease of a red dot finder for rough pointing and the accuracy and light gathering ability of the finderscope for more accurate positioning.

Filters

Filters are other accessories that are worth the money and research time. Sometimes you can find kits that include eyepieces and filters together. Serious astronomers argue that these kits are worthless, but I have found them to be a good way to stock up on decent accessories.

A moon filter is an "important" filter to own, especially if you plan on observing the moon. This filter will enhance some details on the moon's surface (especially if the moon is full and bright). If you have a large telescope (aperture >10in), then a moon filter is almost necessary. The eyepiece image will be very bright and tire your eyes quickly.

There are several different types of colored filters. I will discuss the (1) #25A red and the (2) #80A blue. (1) The #25A red filter can be used to observe details on Mars (polar ice caps, surface detail, dust clouds, etc.) and Jupiter's surface. This filter can also be used to make Venus appear less bright. (2) The #80A blue filter is useful for observing different details on the surface of Jupiter, Saturn, and Mars. The #80A filter blocks less light than the #25A red filter. I use the #80A filter more often because I find that the #25A red filter blocks out too much light. The light transmittance is usually given in the filter's description. The lower the transmittance percentage, the more light it blocks out.

Broadband Light Pollution Reduction (LPR) filters help eliminate some of the man-made lighting that may interfere with some observations. Ultra-High Contrast Nebula (UHC) filters block out even more light than LPR filters. Both of these filters are good filters to experiment with if you have the money. These filters block the "bad" light and let in the "good" light. Cheap LPR and UHC filters will most likely disappoint. It is probably best to put this money towards gas to drive to dark skies instead.

We learned in chapter 3 that every element absorbs and emits different wavelengths of light. If an emission nebula emits wavelengths of light that correspond to the hydrogen atom (which most do), why would we want to include other wavelengths of light? The obvious answer is that we have no choice. The lights that light up our cities also light up our skies. Sodium lighting (a very common type of lighting in and near cities) emits wavelengths of light that correspond to the sodium atom. Unfortunately, if you live near a place that uses this type of lighting, the sodium lighting will interfere with your astronomical observations. LPR and UHC filters block this "sodium light" and allow the "hydrogen light" to pass through. This is basically how these filters work. Sodium lighting is just an example – it is not the only lighting that interferes with astronomical observations. UHC filters block more wavelengths of light than LPR filters.

Figure 6.9 shows three filters: the #80A blue, the #25A red, and a moon filter; from left to right.

Figure 6.9 – Several 1.25" filters and their cases.

Filters usually (if not always) thread into the bottom of an eyepiece. One great thing about filters is that they all use a standard thread (as far as I know). As long as the filter you are purchasing has the correct specified barrel diameter, you are good to go. You should always screw the filter into the part that is closest to your eye. For example, if both your eyepiece and your Barlow lens accept filters, you want to thread the filter into your eyepiece.

The Mount & Tripod

There are a variety of different tripods and mounts a telescope can be attached to. If you are using a truly "cheap" telescope, you have an altazimuth stand. This stand allows for user-controlled up-and-down and side-to-side motion. If you have a somewhat better telescope, you might have slow-motion knobs that allow for precise movements. If you do not have these assists, you will have to do it the old-fashion way. This can get very frustrating (on a side note: if you have not purchased a telescope yet or are looking into buying another – buy a telescope that comes with a stand with slow motion control knobs). Telescopes mounted on altazimuth stands are usually labeled "AZ" telescopes in the description.

Equatorial mounts lock in the declination of the telescope as the Earth rotates. As the night proceeds, an equatorial mount will automatically track an object's declination motion. In other words, you will only have to change the right-ascension (following the Earth's rotation). Telescopes mounted on equatorial mounts are usually labeled "EQ" telescopes in the description.

In astrophotography, a picture may have to be taken over a longer period of time (we will discuss long-exposure photography in chapter 10). Since equatorial mounts automatically control the declination; you will only need one motor to control the right-ascension motion to take long-exposure photographs.

Equatorial mounts are still convenient outside of astrophotography. It is helpful if a telescope can keep an object in the field of view over an extended period of time (especially at higher magnifications). Equatorial mounts are considered the preferred mount in astronomy, but altazimuth mounts can be cheaper and simpler; so, the choice is yours.

If you find yourself with an equatorial mount, you will have to align the mount's right ascension axis with your hemisphere's pole. In the Northern Hemisphere, you can approximate this alignment by simply aligning the axis with Polaris. We will discuss this in detail in chapter 15.

You can still take short-long-exposure (what?) and single-shot photographs using both an altazimuth mount and an equatorial mount without a motor. This will be discussed in chapter 10.

Unfortunately, telescope tripods & mounts may need be serviced often. The bolts on both of my altazimuth tripods need to be tightened occasionally. This is not really that big of a nuisance, just make sure to bring some tools with you in addition to your telescope accessories. Personally, I carry a small tool bag along with me. It holds a case that contains my eyepieces, filters, and 2x Barlow lens; and a couple of the tools needed to tighten the bolts on my telescope. You can never bring too many tools. You do not want to have an observation session ruined by a loose bolt just because you did not bring the wrench along. The most important tools to have are a Phillips head screwdriver, a flat head screwdriver, and an adjustable wrench.

A popular type of mount for larger reflector telescopes is called the Dobsonian mount. Reflector telescopes mounted this way are often called "dobs". Dobsonian telescopes are a joy to use but some people do not enjoy using them. They are by far the cheapest mount for large telescopes.

Altazimuth: Cheapest style of mount. Slow motion controls are necessary for frustration-free sessions.

Equatorial: Slightly more expensive type of mount. Declination and Right-Ascension movements. Motors needed for astrophotography.

Dobsonian: Type of mount larger reflector telescopes are typically mounted on. Best option for relaxed visual use.

For the budget astronomer...

DSO Astrophotography: Get an equatorial mount or a camera star tracker. I have used a 50mm f/5 mounted on a star tracker that I used for my DSO astrophotography pursuits. If you want to use a telescope larger than a 50mm f/5, I would suggest spending the big bucks and getting a high quality computerized equatorial mount. You cannot spend too much money, but you will almost certainly not spend enough the first time. Buy it now & keep it forever. Remember: "buy once, cry once"!

Planetary/Lunar Astrophotography or Visual use: For visual, just get a dobsonian. You may find its design awkward, but this is your best value. For lunar/planetary, equatorial works but computerized AZ is easier. I have motorized a 12" dobsonian telescope for lunar/planetary imaging. A fork mounted SCT is very simple and easy to use if you think the dobsonian style may be awkward. I have a 10" SCT that I put on an equatorial mount for lunar and planetary imaging, but I already had a quality equatorial mount laying around from my DSO astrophotography efforts.

In summary, get a *nice* equatorial mount. My equatorial mount cost more than I paid for my current car. They are important. You can put little telescopes on it for DSO imaging and switch to larger ones for planetary/lunar imaging and even visual astronomy. Equatorial mounts are worth the investment once you have proved to yourself that you love this hobby.

One last word on tripods... If you are imaging DSOs or putting a heavy load on your mount, a good 2" tripod is probably the minimum. I have switched to a portable pier and it is worth the extra transportation hassle. If you do not have the money to upgrade, avoid extending the legs on your smaller tripods (keep the legs fully collapsed).

Chapter 7

Improving a Cheap Telescope

Perhaps the best "improvement" for a cheap telescope is simply maintaining it. Temperature and dust are your biggest enemies.

Cold skies are usually clear skies, but along with the cold comes dew. You should generally avoid taking a cheap telescope outside in below freezing temperatures especially if it is humid outside. Excessive cold can damage the plastic and optics on a cheap telescope.

That being said, I have taken my 50mm refractor telescope outside at temperatures well below freezing and have survived. You can take a telescope out into the excessive cold, but just be careful and limit the amount of time you spend outside.

The part of the telescope that extends from and surrounds the objective lens is called the dew cap, and it is there to collect dew before the objective lens does. If you do see water on your objective lens, you can do a few things, but you cannot wipe the objective lens.

DO NOT WIPE THE OBJECTIVE LENS!!!

I do not know how to make that any clearer. Wiping the objective lens will ruin the quality of the optics, so do not touch it. If you see water on your objective lens, wrap a plastic bag around the objective lens potion of the telescope and make sure it is sealed. Let it sit for a few minutes and then bring it inside. The water will collect in the bag and not on the objective lens. While you are outside in the cold, do not accidentally breathe on the eyepiece. You should not breathe on any optics anyway, but especially avoid doing this in the cold (the water vapor in your breath will condense onto the cold optics).

Always keep the caps on your optics. Keep the objective cap on unless you are using it. Keep your eyepieces covered. You want to minimize the amount of time your optics see anything other than the caps that keep them safe. If you get dust on your optics, or a fingerprint, or anything else; do not attempt to clean.

DO NOT CLEAN YOUR OPTICS!!!

Again, I do not know how to make myself any clearer. Do not clean them. If you *must* do something, you have a few options. Like all hobbies, there are people who are very serious about astronomy and people who do it for fun. You know who you are. I am assuming you are doing it for fun since you bought a $15 book. Your cheap optics can be cleaned in a variety of ways.

(1) Get one of those cans that blow air. The ones designed for cleaning electrical equipment and circuit boards. Blow the dust off your optics. Make sure the air is compressed air and does not contain any harsh chemicals. These cans are not easily found.

(2) Get one of those manual air blowers. I am not sure what to call them. Some are called "manual air blowers", others "squeeze dusters", and others "bulb dusters". They are essentially a little ball or bulb you can squeeze and air blows out from a small opening. These are cheap and I like using these to blow the dust off optics. These are easily found.

(3) You can try to wipe your optics with a microfiber cloth or optics cleaning pen. I have never used either of these, but if your optics are cheap and you are just doing this for fun, go ahead and try it (just be gentle). Maybe I am over-thinking this…

Avoid the excessive cold, keep your caps on, and do not clean the optics. If you must clean them, avoid doing so often.

Another easy "improvement" is using your telescope properly. You want your *entire* telescope to be in thermal equilibrium with its surroundings (refraction of the light will occur when there is a temperature difference and this will distort the image). For a small telescope, this is really quite simple. Place your telescope outside for a few minutes before using it. When you get to it, it will be ready. This is an advantage associated with smaller telescopes. Smaller refractor telescopes reach thermal equilibrium much faster than larger reflector telescopes so enjoy one of the few advantages you have.

You are also going to want your eyes to adjust to the dark. You will be surprised how big of a difference that can make. You will be able see more faint objects and details after your eyes have adjusted to the complete darkness for ~10 minutes or so. If you *need* light while observing, most people use a small red flashlight as this will not ruin any progress your eyes have made adjusting to the dark. Simply apply red paint to the front of any small flashlight you have at home, wrap red cellophane around the front of a flashlight, or buy a red flashlight.

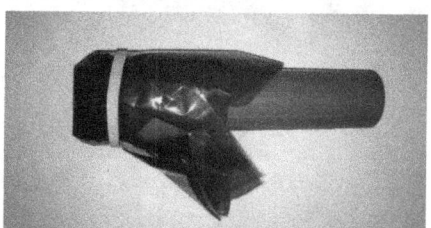

Figure 7.1 – Red cellophane wrapped around the front of a flashlight. Use a rubber band to secure the cellophane. You may have to use a few layers of the cellophane.

We will talk more about this shortly, but I have some room on this page. For best results, you should use your telescope during clear, steady, and dark skies. Clear skies meaning no clouds and good transparency, steady skies meaning a steady atmosphere, and dark skies meaning away from light pollution.

The rack and pinion gear that controls the focus is almost always too lose or too tight. You can tighten or loosen the screws on the focuser however you like. You can see these screws in figure 7.2 under the shaft attached to the knobs. The procedure may be different for your telescope. My 90mm refractor has a thumb screw that controls how tight or lose the focus mechanism is.

Figure 7.2 – Screws for focus mechanism.

If the screws are too tight, the telescope might shake excessively while you focus. This makes it difficult to tell when an object is actually focused. If the screws are too loose, gravity might change the focus for you or the image might be distorted in some way. You can experiment with this and find what you prefer.

The tripod you received with your cheap telescope is probably too light and unsteady. I find that the best improvements for these stands (and the telescopes they hold) is to weigh them down. I use a tool bag filled with the heaviest tools I can find and tie it around the top of the stand (where the telescope and stand are connected). The tool bag should be suspended in the air when the legs are extended, and this added weight will greatly improve the steadiness (see figure 7.3). This modification is almost necessary for lighter, smaller aperture telescopes. I have also seen some people use water jugs as the weight, but since you should be carrying around a tool bag anyway, you might as well use it to weigh down the stand. My 90mm refractor is heavy enough so I do not have to do this, but I do need to do this with my 50mm.

Figure 7.3 – Weighing down the stand.

Fans on reflectors:

If you have a reflector larger than 8 inches, fans are a good idea. A fan on the primary mirror will cool the telescope faster and keep it at approximately ambient temperature throughout the night, possibly preventing dew. I use 12 volt computer fans.

A fan can be placed underneath the primary mirror at the bottom of the reflector tube. Usually, the fan blows air on the primary but some people have it draw air from the tube and blow it out the bottom. Each have their own advantages so experiment and see what works best for your telescope. I have even put a fan on a secondary mirror to keep dew off!

Dew control using resistors:

Please be careful and read more about this before you attempt it on your own!

Simply put, resistors consume electricity and produce heat. If you wrap a strip of resistors around the dew cap of a refractor or the secondary of a reflector, you can quickly remove dew from your optics. Dew on the primary mirror of a reflector is generally not that great of a concern especially if you have a solid tube reflector.

You probably only need 1-3 watts of power.

$$resistance\ needed = \frac{Volts^2}{Watts}$$

You need to be very careful to ensure there can be no shorts or bad connections. If you do not feel comfortable with this, a low power hair dryer should be just fine!

They also sell 12V dew strips for this purpose. The choice is yours.

Replacing the focuser

On cheap achromat refractors, the focusers can be very bad. Cheap achromats are valuable for affordable narrowband imaging if you re-focus for every filter. Usually the thing stopping people is a bad focuser. They can usually be upgraded easily.

On SCT telescopes I am a huge fan of adding rear-mounted crayford focusers. They not only adapt your telescope to a 2" visual back, but their very presence is better for lunar & planetary imaging. The reason is two-fold. (1) It is easier to achieve fine focus with a 2-speed crayford focuser than the focus knob that moves the primary. (2) Your collimation will be better.

You can collimate after moving the primary mirror with the stock focus knob to a rough focus position. You can do a star test by using the crayford focuser to take a star slightly out of focus. You can then achieve fine focus with the crayford without disturbing fine collimation. I am a firm believer in that when you adjust focus with the stock focus knob and move the primary, your collimation is ruined.

Take pride in your collimation

Collimation is brutally important with reflectors and SCTs. Don't ignore it. Rather, embrace collimation and have fun with it. Good collimation with ensure a much better experience. We will discuss collimation more in later chapters.

Achieving better focus

Focusing is tough for beginners in both visual astronomy & astrophotography. Visual is easy. Adjust focus until you see a very dim big circle. Keep going in the same direction as the star get smaller and smaller. Eventually it will pass a point where it is the smallest and then it starts to get bigger again. We are interested in getting the smallest dot possible. Stars are not discs with our telescopes – they are points. The following image represents a time profile of focus as you constantly spin the focuser knob clockwise.

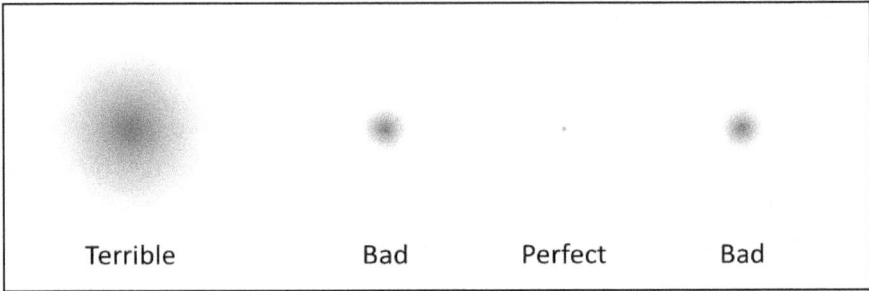

Perfect focus is difficult to achieve.

For DSO photography, focus is much more important. A Bahtinov mask works so well for DSO imaging that I would require it if I could.

For lunar and planetary, the best practice is to focus on the smallest detail you can see on the computer screen and adjust focus back & forth until that smallest detail looks the sharpest. It can get frustrating, but that is part of the fun. A motorized focuser cuts down on the shaking and makes this form of imaging much more enjoyable.

The Coma Corrector

For visual use with a Newtonian faster than F/5, you are likely going to want a coma corrector to fix comet-shaped stars near the edges of your field of view. The coma corrector setting will depend on the eyepiece you are using, so experimentation is key. Coma will be more obvious with longer focal length eyepieces with larger fields of view. Depending on your observation style, you may not need a coma corrector at all.

For DSO imaging with a Newtonian faster than F/7, you are likely going to need a coma corrector. Ideally get one without spherical aberrations and one that can correct for the entire size of your camera's sensor. For example, if you have a 22mm sensor diagonal, look for a coma corrector that can correct for a 22mm image circle.

The coma corrector spacing you need depends on your optical system. Even if the manufacturer recommends 55mm backspacing, please experiment with that value. Most of the time, manufacturers determine coma correct backspacing specs by using longer focal length instruments. I image with a 6" F5 Newtonian and have discovered that the backspacing I have to use can be anywhere from 0-4mm different than what the manufacturer recommends.

For lunar & planetary imaging, you do not need a coma corrector.

The Field Flattener

For imaging DSOs only, a field flattener will be needed with most refractors, depending on your sensor size. Again, make sure any field flattener you purchase can handle the image circle required by your sensor. Also again, you may need to play with backspacing.

Backspacing

A coma corrector and field flattener will both come with a manufacturer's recommended backspacing. Please, plan to experiment with this value.

I find the best way to experiment with this value is to get a variable T2 or M48 spacer adapter. These allow you to fine tune the spacing between your camera's sensor and the corrector lens assembly within 0.5mm increments. Backspacing is important, do not ignore it. The good news is once you figure it out, you can set it for life and never have to experiment again. Small changes can have big impact on your image quality.

For example, the image on the left was taken with my 6" F/4 Newtonian with a coma corrector with 55mm backspacing. The second image was taken with the backspacing adjusted to 56mm.

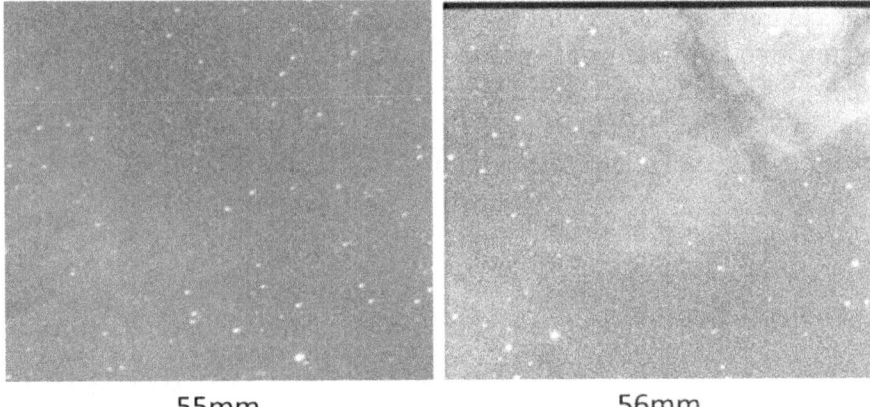

| 55mm | 56mm |

The difference is quite substantial for 1mm difference. These are screenshots of the top-right most section of two different photos. Notice how the stars look like comets with 55m backspacing and look rounder with 56mm backspacing. The manufacturer recommended 55mm.

Insulating a SC or Maksutov telescopes

As a chemical engineer by trade, I could talk about heat transfer all day. This is, for some unknown reason, a heavily debated issue in amateur astronomy. It is almost funny.

The problem is closed tube telescopes are sealed on both ends. The air is trapped between the optical surfaces. If the air trapped between both surfaces is not at thermal equilibrium, the thermal currents will distort the image. This is especially true if the telescope objective is greater than 6 inches. The interior air needs to be very steady since these telescopes force light to pass through the interior air three (3) times.

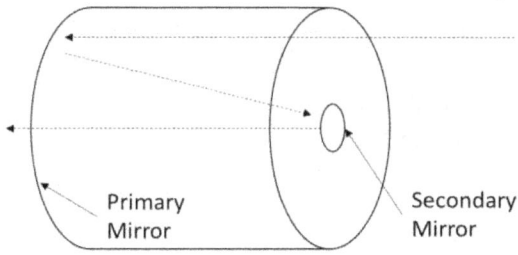

Primary Mirror   Secondary Mirror

If the inside air is unsteady, the resultant eyepiece or camera image will be unsteady. This much is agreed upon by everyone. The question remains, how can we prevent the interior air from being unsteady?

Temperature is the problem. If there is hot air & cold air inside the telescope tube, the air will mix. The mixing air will distort the image the telescope produces. We need to keep the temperature inside the tube constant to prevent this mixing and image distortion. This "mixing" is referred to as thermal currents.

If you store your telescope indoors (say, 70F) and bring it outside (say, 32F), the outside air will begin to cool down the telescope internals (trapped air & mirrors). This "cooling down" takes a very long time, mostly since the glass mirrors shed residual temperature very slowly. During the "cool down" the interior telescope air is mixing and distorting the image. Therefore, large versions of these telescopes cannot be used immediately upon bringing outdoors if they are stored indoors at much warmer temperatures.

Some people say that the telescopes cool down on their own if you just sit them outside for a few hours. Some people say they need active cooling (via fans & vents). Some people only need passive cooling (via vents). Some people put ice packs on the telescope tube walls to cool them down faster. Some people insulate the telescopes. It appears that the correct "fix" is location dependent.

There is a lot of information about this topic online. While entertaining and genuinely informative, it can be confusing. As a chemical engineer by day, I decided to look at the insulation a little closer. There is a lot of talk about a "well insulated" telescope but I don't see people often refer to any specifications. So, let's waste some time over-analyzing the insulation to see if we can get a better idea of what a "well insulated" SC or Maksutov telescope is.

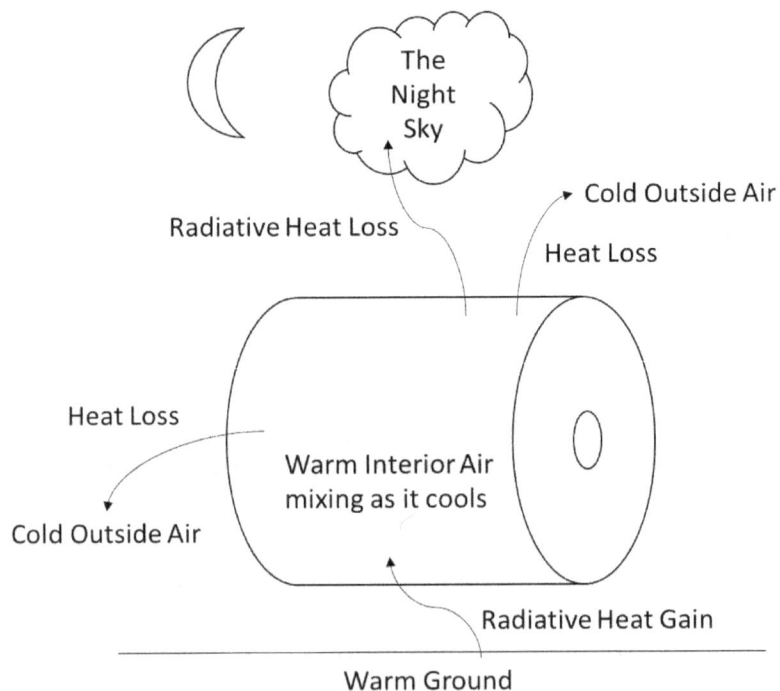

The previous graphic highlights how much heat transfer is going on.

The radiative heat loss & gain is solved easily using a reflective surface. Reflective bubble-wrap insulation is a very popular way of achieving this. We will finish any insulation with a layer of this reflective insulation to take care of the radiative heat loss and gain. We will ignore radiative heat loss or gain in any calculations.

After insulating, the heat transfer path becomes severely impaired. This is good. If we insulate sufficiently, it may be able to keep the telescope air stable enough for immediate observation, even without an initial cooldown.

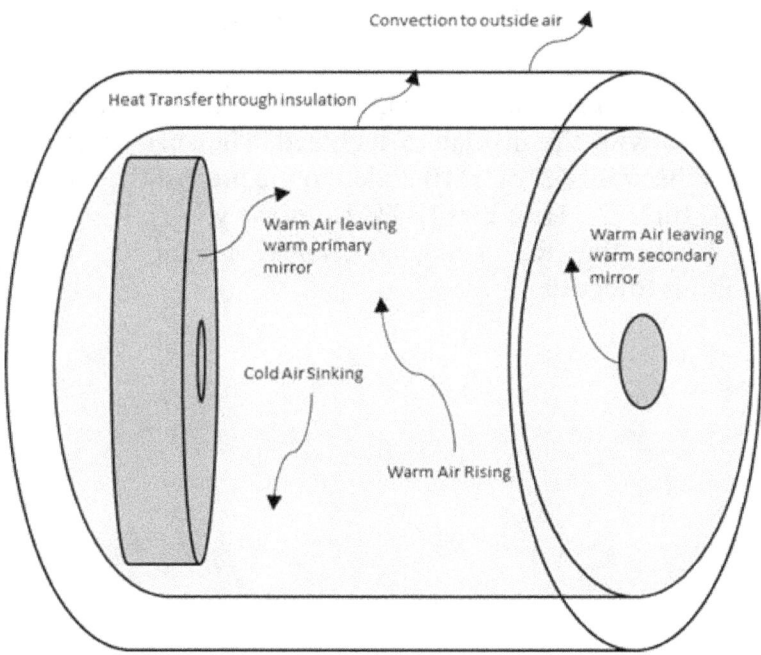

If we can slow the heat transfer through the insulation enough, perhaps the thermal currents inside of the telescope tube will be eliminated. Obviously, the best way to do this is the choose the proper insulation material and insulation layer thickness.

This is not going to be a brutally accurate model; this is not worth the effort it would take to do that. Instead, we are just exploring what insulation materials and thicknesses are necessary to sufficiently slow the creation of thermal currents inside of the telescope tube.

I am going to approximate the telescope to be a 11.5" OD and 20" long pipe. I am going to ignore the front corrector plate and rear. I know that will drive people crazy. Again, we are just looking to get an understanding of what is necessary.

First, we can do an incredibly rough analysis. Let's treat it like a house.

$$Heat\ Loss\left(\frac{BTU}{hr}\right) = \frac{Surface\ area * (T\ difference)}{Rvalue}$$

Let's use 1" of insulation layer thickness so we can calculate the surface area with the insulation included. The surface area is the exterior tube walls (2*pi*r*L) added to the area of the back of the telescope (pi*r^2). Heat loss (BTU/hr) is the y-axis. R-value is the x-axis. Looks like R2 is a good choice? Looks like low-level insulation is effective.

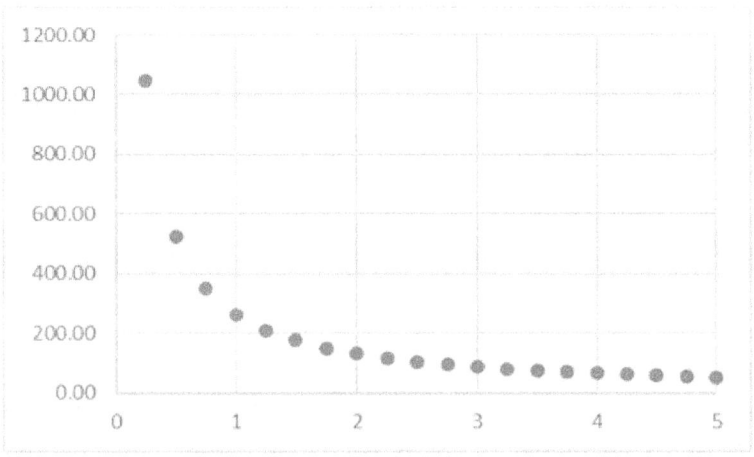

R-value	Initial Telescope Temperature					
	T=80F	T=70F	T=60F	T=50F	T=40F	T=34F
1.25	322.5368	255.3417	188.1465	120.9513	53.75614	13.43904
2	201.5855	159.5885	117.5916	75.59457	33.59759	8.399397
3	134.3904	106.3924	78.39437	50.39638	22.39839	5.599598
4	100.7928	79.79427	58.79578	37.79729	16.79879	4.199698
5	80.63421	63.83542	47.03662	30.23783	13.43904	3.359759
	Heat Loss Rate (BTU/Hr)					

It also looks like initial telescope temperature may play a bigger role than insulation, even at R4, which is roughly a full inch of quality, industrial polyurethane insulation. This is not necessarily promising, unless a heat loss rate of ~60BTU/hr is acceptable and doesn't disturb the image too much.

Uninsulated Telescope

The heat transfer pathways we are considering are:
- Convection inside the telescope tube (convection1)
- Conduction through telescope tube wall (conduction)
- Forced & free convection to outside air (convection2)

Convection resistance inside the tube wall is easy, it is 1/h. "h" here is the heat transfer coefficient. 5W/m2K is the lower end of the standard value for free convection of atmospheric pressure gas (5-35), so let's use that since this will also account for the heat loss of the mirrors, which will be very slow. I later adjusted this value to 2 to better match real life cool-down times.

Resistance of conduction through the tube wall is simple. This resistance is:

$$R_{conduction} = \frac{D3 * \ln(\frac{D2}{D1})}{2k}$$

Where D3 is the insulation OD, D2 is the tube OD, and D1 is the tube ID. Here, D3 and D2 are equal since there is no insulation yet.

117

For steel, we will use a thermal conductivity of 45 W/mK ignoring dependence on temperature since our model is a very basic approximation.

Convection resistance to outside air is a very complicated term. For forced convection, we use the Churchill and Bernstein equation for Nusselt number.

$$Nu_{forced} = 0.3 + \frac{0.62 Re^{\frac{1}{2}} Pr^{\frac{1}{3}}}{\left[1 + (\frac{0.4}{Pr})^{\frac{2}{3}}\right]^{\frac{1}{4}}} \left(1 + (\frac{Re}{282000})^{\frac{5}{8}}\right)^{\frac{4}{5}}$$

"Re" factors in wind speed, which I used ~2mph for. For Pr I just used a literature value of 0.714. For free convection, we use the Churchill and Chu equation for Nusselt Number:

$$Nu_{free} = \left[0.6 + \frac{0.387 Ra^{\frac{1}{6}}}{\left[1 + \left(\frac{0.559}{Pr}\right)^{\frac{9}{16}}\right]^{\frac{8}{27}}}\right]^{2}$$

The same Pr number was used. The forced and free convection heat transfer coefficients were calculated as follows:

$$h_{forced} = \frac{Nu_{forced} * k_{air}}{D3}$$

$$h_{free} = \frac{Nu_{free} * k_{air}}{D3}$$

The combined heat transfer coefficient was calculated as follows:

$$Nu_{combined} = \left(Nu_{forced}{}^4 + Nu_{free}{}^4\right)^{\frac{1}{4}}$$

$$h_{combined} = \frac{Nu_{combined} * k_{air}}{D3}$$

The overall external convection resistance was calculated as follows:

$$R_{convection\ 2} = \frac{1}{h_{combined}}$$

I ignored several fine details of this calculation for simplicity, but it is easy to calculate in excel. The overall resistance is the sum of convection1, conduction, and convection2. The cooling power per meter of "pipe" or telescope tube is:

$$Power\left(\frac{W}{m}\right) = \frac{Temperature\ Change}{R} * 3.1415 * D3$$

Our tube is only 20" long, so we just have to multiply the cooling power by 0.508m to calculate that the cooling power of the uninsulated tube is ~15.25W if the temperature differential is 21C. It is only ~7.23W if the temperature differential is 10C.

Insulated Telescope

The heat transfer pathways we are considering are:
- Convection inside the telescope tube (convection1)
- Conduction through telescope tube wall (conduction1)
- Conduction through the insulation (conduction2)
- Forced & free convection to outside air (convection2)

Convection resistance inside the tube wall same as before. 2W/m2K.

Resistance of conduction through the tube wall same equation as before, but this time, D3 and D2 will be different.

Convection resistance to outside air same complex story, with minor differences we won't discuss but are easy to capture in an excel spreadsheet.

Resistance of conduction through the insulation is easy to calculate using the following equation:

$$R_{conduction\ 2} = \frac{D3 * \ln(\frac{D3}{D2})}{2k}$$

"k" is the thermal conductivity and is dependent on the insulation material. It seems the best option for telescope insulators is quality, industrial polyurethane foam insulation. A generous thermal conductivity value for that insulation is 0.035W/mK at the temperatures we see.

The overall resistance is the sum of convection1, conduction1, conduction2, and convection2. The cooling power of the insulated tube is ~8W if the temperature differential is 21C. It is only ~3.82W if the temperature differential is 10C. This assumes 1" insulation thickness was used.

Heat Lost Calculation

We can calculate how much energy was given up by the internal mirrors and air to the surroundings via basic density, volume, and mass calculations. We know the internal mirrors are borosilicate and this helps us select the appropriate specific heat and density values. We can take a guess about the internal mirrors' dimensions to get their volume.

$$Heat\ Lost\ (J) = \frac{Temperature\ Differential\ (degC)}{V_{air} * D_{air} * SH_{air} + V_{glass} * D_{glass} * SH_{glass}}$$

Where V is volume, D is density, and SH is specific heat. For a 10C temperature differential, 43542J of energy from the internal air and mirrors is given up to the outside air.

An incredibly rough estimation can now be made for total cool down time from storage temperature to ambient:

$$Cool\ down\ time\ (s) = \frac{Cooling\ Power\ (W)}{Heat\ Lost\ (J)}$$

Temperature Profile Calculation

The temperature drop throughout the cooldown is not linear, but we can simplify the profile by using the theory of exponential decay. In order to calculate the time-constant, we need to calculate the cooling rate:

$$Cooling\ Rate\ \left(\frac{degC}{hr}\right) = \frac{Cooling\ Power\ \left(\frac{J}{s}\right) * 3600\frac{s}{hr}}{(V_{air} * D_{air} * SH_{air} + V_{glass} * D_{glass} * SH_{glass})}$$

The time constant can be calculated as follows:

$$tau(hours) = \frac{Temperature\ Differential\ (TD)\ (degC)}{Cooling\ Rate\ \left(\frac{DegC}{hr}\right)}$$

The temperature profile of the telescope internals during the cool down can be calculated per the following equation:

$$Temperature = Final\ temperature(degC) + TD(degC) * e^{\left(\frac{time\ (hours)}{tau\ (hours)}\right)}$$

Temperature Profiles

Consider that the telescope is stored indoors at 21C (~70F) and is brought outside where the temperature is 0C (~32F). Also consider the case that the telescope starting temperature is 10C. Here are the temperature profiles for various insulation layer thicknesses.

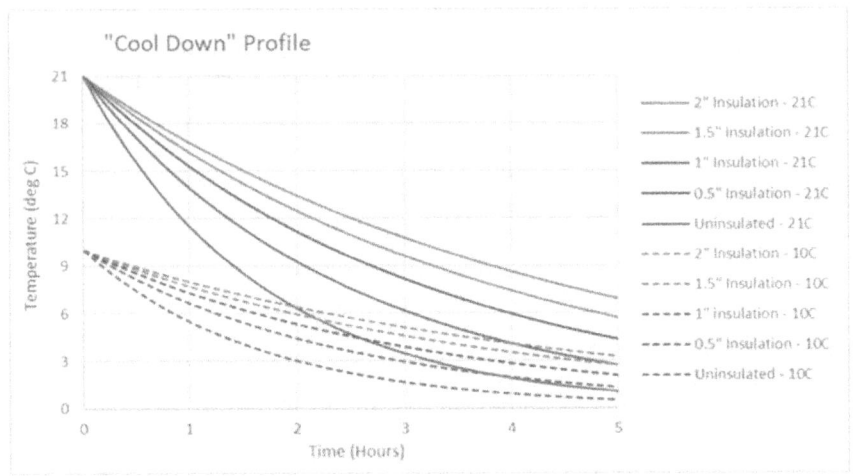

After doing all this math, it does not look like conduction & convection are major problems since even heavy insulation doesn't have much of an effect. The best solution is probably a thin layer of insulation (~R1) and reflective insulation. It may very well be that radiative heat loss and gain are the bigger problems.

Another trick that I like to implement is to point the telescope toward the ground so that the focuser sticks up towards the sky and let the telescope cool this way. It is best to place a filter media of some sort over the focuser and remove the focuser cap. This will allow the hot air inside the telescope to rise and escape the telescope. It is easier to keep a telescope insulated if the temperature differential between the trapped air in the surrounding air is not so great.

(Astrophotography) cable management

While not a direct improvement to the ability of the telescope to reveal the cosmos in greater detail, cable management can help astrophotography sessions be cleaner and easier to manage. It can even help visual setups if you have cables laying around. Tripping hazards are real, despite how funny that sounds. Unplugged cables can ruin a clear night.

Cable management can be very easy with just the use of cable ties and some creativity. It can also be very fun if you enjoy this sort of thing. I have re-done my cable management several times, each with minor improvements. I photographed my last cable management session. I will include the photos below.

This is obviously something that is very equipment-specific, but I just want to show one example of how I did it. Maybe it will spark an idea or two for you.

Before I reveal the photos (you must be on the edge of your seat), I want to share a cable tie trick that I do not think many people are aware of.

First, take your cables and line them next to each other. For this example, let's consider 4 cables. Second, wrap all of them loosely with one cable tie.

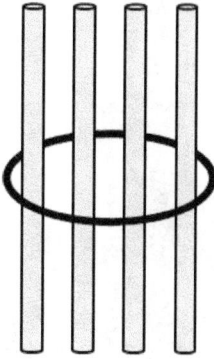

Count the number of spaces between the wires. This is the number of additional cable ties you will need. Grab that number of cable ties and feed them between the wires and around the main cable tie as shown.

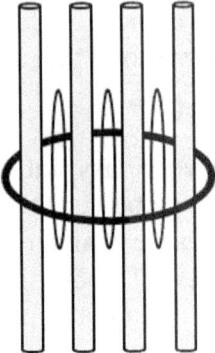

Finally, tighten up all the cable ties. This is a great way to manage cables! Here is a photo of this technique used to manage the flow of 3 cables.

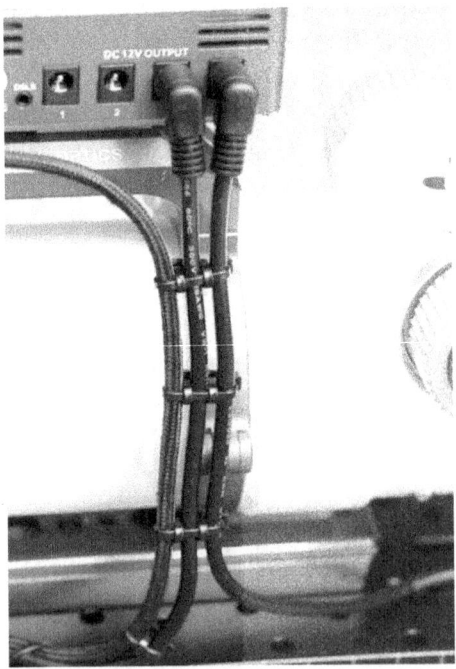

As promised, here are some photos of cable management for an astrophotography setup.

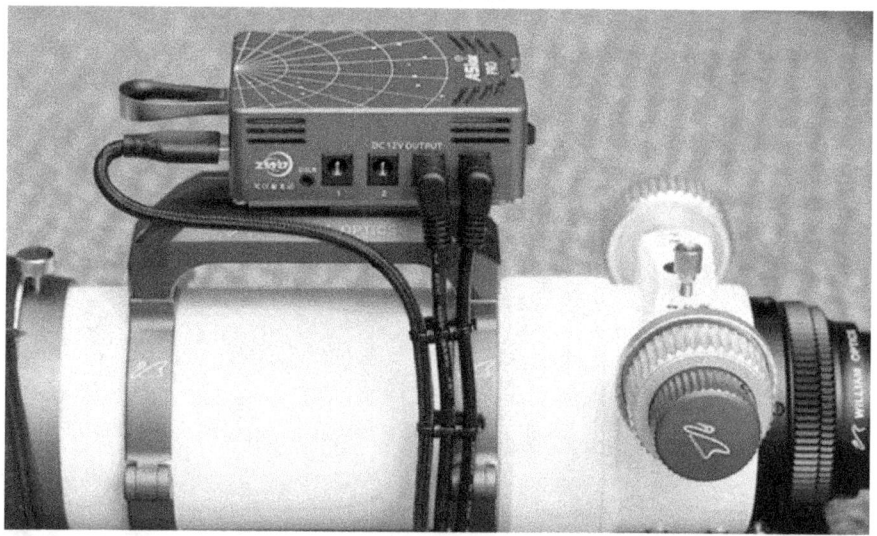

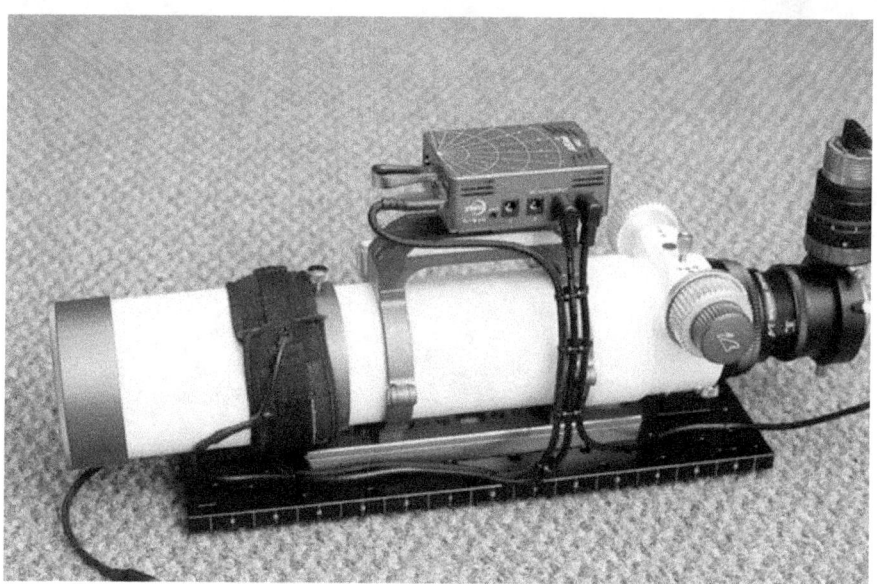

I really love the way this turned out. It makes setup feel easier and more stable.

This telescope's dovetail is actually mounted on a larger dovetail for stability reasons. Here, I used cable ties to attach cables to this larger dovetail.

Part 3: Observations & Astrophotography

There is no doubt in my mind that you are going to want to observe the sky when it is cold outside. There are many objects that are only visible during the colder months, like the constellation Orion (in the Northern Hemisphere). We have already discussed how *too* cold can be bad for the telescope, but cold, in general, can be bad for *you*. Humans do not generally like the cold weather, so dress warm and be patient. Rushing and getting angry is not going to help anything. Telescopes will generally survive the cold and so can you if you dress warm.

Observing deep sky objects is an easier task when the skies are dark. The moon also contributes to the darkness of the sky at night. If the moon is *not* up, the skies will be darker and fainter objects will be easier to see; if the moon is up, the skies will be brighter and fainter objects will be harder to see. As a beginner using a 50mm refractor near the city of Philadelphia, I found it nearly impossible to find deep sky objects when the moon was up. You can still view multiple star systems and planets when the moon is up, although you may find that these observations are also better when the moon is not in the sky.

It is generally better to observe something when it is highest in the sky, directly above you. A small telescope might not be able to tell the difference but it is still the best practice. The light coming from whatever you are observing has to travel through atmosphere in order to get to your telescope, and minimizing the amount of atmosphere it has to travel through will increase the steadiness and quality of the image. When an object is directly overhead, the light coming from the object has to travel through the least amount of atmosphere (see the figure on the next page).

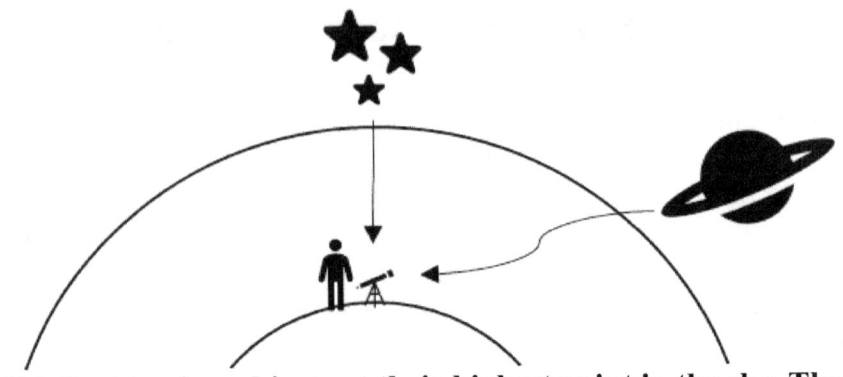

It is best to view objects at their highest point in the sky. The dome is the Earth's atmosphere.

In the figure, the triangle is directly above the observer on Earth. The light from the triangle, therefore, has to travel through the least amount of atmosphere. The two squares, however, are located near the horizon. The light coming from the squares, therefore, has to travel through more atmosphere. The triangle is the best object to observe at this time.

All objects rise in the east and set in the west, just like the sun. You may wish to observe objects in the west first since they are the first to set and disappear. Stars and deep sky objects circle around the Earth's axis. The sky changes with the seasons. The stars and deep sky objects visible in the summer night sky will be different than the stars and deep sky objects visible in the winter night sky. When you are observing, you typically want to put your telescope on soft ground (grass). If your telescope is on hard ground, it is susceptible to the ground's vibrations. Softer ground is able to absorb these vibrations.

We are going to discuss how to observe the moon, DSOs, the planets, and the sun in chapters 8 and 9. A quick summary (if you are in a hurry) can be found on the following page.

The Moon: You are free to try any magnification you want. Experiment with high and low magnifications. Experiment with filters. There are no rules when it comes to observing the moon. Just do what you enjoy the most (I like to use high magnifications and focus on particular craters and regions). Observe the moon whenever it is up.

Deep Sky Objects (DSOs): You are actually going to want to use low-to-medium magnifications (generally less than 100 times) since these objects are fairly large. Observations are best when the sky is dark and steady. If you are in a light-polluted area, you may want to consider LPR or UHC filters. The larger the aperture, the better. DSOs can be frustrating to find and observe. DSOs are also difficult to take pictures of and usually require long-exposure photography.

Planets: Planets look like bright stars in the night sky so they are relatively easy to find. Get a high magnification and you are good to go. Magnifications of at least 100 times are recommended, but magnifications over 200 times are not generally necessary. Colored filters show different details on the surfaces of some planets and they should be experimented with. Planets are, generally, much easier to find and take pictures of.

The Sun:

Again, I repeat my warning:

<u>WARNING</u>: Do <u>NOT</u> look at the sun with your telescope. You will damage your eyes (and the telescope's optics) unless you take the necessary precautions. Do not observe the sun until you have done your research and have all the appropriate equipment.

In order to safely observe the sun, you need to use proper solar filters. If you do it safely, observing the sun can be rewarding.

Dark Skies

Light-pollution greatly affects astronomical observations using a telescope and so does the moon. If you live in a light-polluted area, you can still observe a good amount of the sky with a telescope. That being said, those same observations would be much better in darker skies.

Light-pollution maps show you which areas are effected the most by light-pollution. I find that these maps are somewhat inaccurate, but still give a rough sense of where you want to go to try to find dark skies. Dark sky parks are areas with especially (you guessed it) dark skies. Once you get used to the hobby and can find a couple deep sky objects on your own, you should plan a trip to a dark sky park, or any area less effected by light-pollution.

If you do not live close to a dark sky park or any less light-polluted areas, try to find a dark park. If the park has hills you can get on top of, it is even better. The rule is simple: the darker the skies, the better the observations. That being said, it would not be wise to plan a trip to a dark sky park if the moon would be up at that time. Go outside during a full moon and go outside during a new moon. The skies are much darker during the new moon.

Steady Skies

Not only do you need dark skies, you need steady skies too. You cannot control the steadiness of the atmosphere, so there is really nothing you can do about this. All you can do is wait for a steady night. The rule is simple: the steadier the skies, the better the observations will be and the higher the magnifications you can use. Some astronomy and weather sites list the seeing conditions for a given region on a given night on a scale of 1 to 5 (5 being the best and 1 being the worst).

Oh, you need the skies to be clear too. Clouds hide the night sky.

Chapter 8

Observing the Moon

The moon is absolutely the best thing to observe with a cheap telescope, but a very small portion of the population actually understands the moon.

The moon is *not* only visible at night. You may have noticed the moon when the sun was up and been confused, but this is normal. The moon does not magically rise when the sun sets and set when the sun rises. The moon rises in the east and sets in the west, just like the sun, but the relative positions of the Earth, sun, and moon determine when the moon rises and what it looks like when it does.

We should all be familiar with the full moon. We should all be familiar with the new moon. After the full moon, the moon slowly transforms into the new moon, changing the way it looks it process. After the new moon, the moon slowly transforms into the full moon, changing the way it looks it process. When the moon is transitioning from a full moon to a new moon, it is considered a "waning" moon. When the moon is transitioning from a new moon to a full moon, it is considered a "waxing" moon.

FULL → NEW (waning moon)

NEW → FULL (waxing moon)

Let's start at the new moon. After the new moon, the moon is considered a waxing crescent moon. The first quarter moon occurs after the waxing crescent moon. After the first quarter moon, the moon is considered a waxing gibbous moon. The full moon occurs after the waxing gibbous moon. After the full moon, the moon is considered a waning gibbous moon. The last quarter moon occurs after the waning gibbous moon. After the last quarter moon, the moon is considered a waning crescent moon. The new moon occurs after the waning crescent moon. *(new moon, waxing crescent, first quarter, waxing gibbous, full, waning gibbous, last quarter, waning crescent, new moon again).*

As far as observing goes, use your telescope to look at the moon whenever it is up. Finding other objects will be very difficult in light polluted areas with a bright moon. It is best to save other observations for nights when the moon is not up. I have been repeating this so that you remember it!

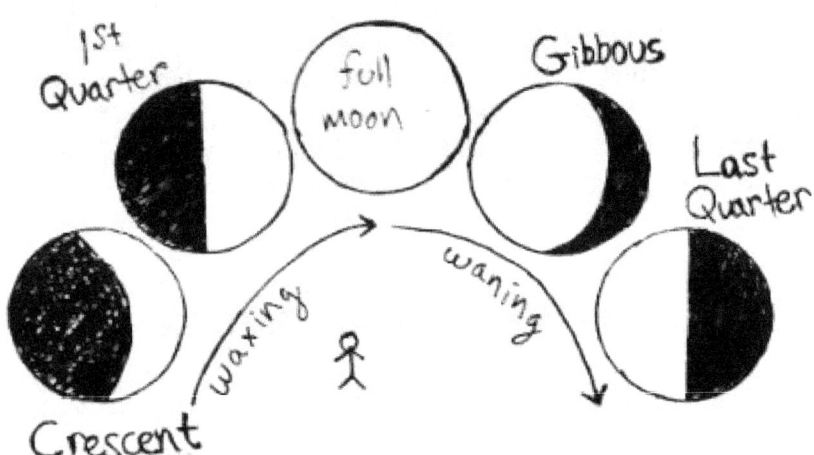

Figure 8.1 – The phases of the moon.

The moon rises and sets at different times during its lunar cycle. The waxing crescent moon sets in the west soon after the sun. It disappears from the night sky early in the night. As it gets closer to the first quarter moon, it sets later in the night, so it is up longer. The first quarter moon is at its highest point in the sky at sunset and sets around midnight. Therefore, you are able to observe faint objects after midnight.

The waxing gibbous moon reaches its highest point in the sky closer to midnight. When the sun sets the waxing gibbous moon is closer to the eastern horizon, so it is up for longer in the night. The full moon rises when the sun sets and sets when the sun rises. The full moon is at its highest point in the sky at midnight. The full moon is up the entire night and makes observing faint objects very difficult.

The waning gibbous moon rises later in the night and reaches its highest point after midnight. It is still up when the sun rises, but it is almost setting. The last quarter moon rises around midnight, so you are able to observe faint objects before midnight.

The waning crescent moon rises later in the night as it approaches the new moon. The new moon rises at sunrise and sets at sunset. You are free to observe faint objects all night because the new moon will not be in the night sky.

If this seems complicated, it will make more sense once you start paying attention to the moon. The changes from night to night are actually quite drastic, maybe more so than you would think. The left side of the moon is lit during the waning moon and the right side of the moon is lit during the waxing moon.

Moon Phase	Rise Time	Set Time
New Moon	~7am	~7pm
Early Waxing Crescent	~9am	~9pm
Late Waxing Crescent	~11am	~11pm
First Quarter	~12pm	~12am
Early Waxing Gibbous	~3pm	~3am
Late Waxing Gibbous	~6pm	~6am
Full Moon	~7pm	~7am
Early Waning Gibbous	~9pm	~9am
Late Waning Gibbous	~11pm	~11am
Last Quarter	~12am	~12pm
Early Waning Crescent	~3am	~3pm
Late Waning Crescent	~6am	~6pm

This table is meant to give you a rough sense of when certain moon phases rise and set. This will let you predict when the moon is going to be up, so that you can plan your observations accordingly. Moon calendars are relatively easy to find. In fact, your calendar may already include the lunar phases.

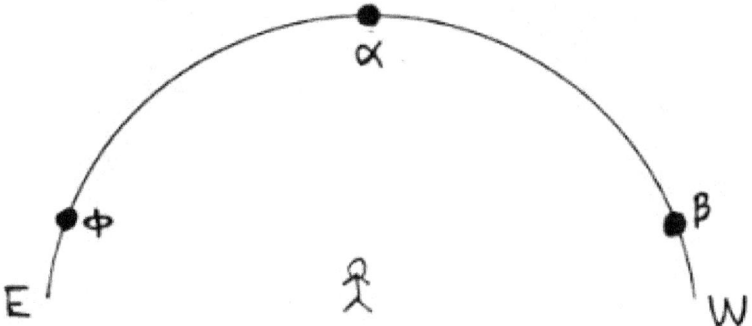

Figure 8.2 – The positions of the moon at sunset – β represents the waxing crescent moon, α represents the first quarter moon, and Φ represents the waning gibbous moon (assuming the sun is at W).

The moon has craters and maria (mare is the singular). Maria are the large, dark spots on the moon. Figure 8.3 shows the main Maria and the included table lists their names. This can be found on the next page so that they could be kept together.

The crater at the bottom of figure 8.3 is the crater Tycho. Tycho is a very young crater (only ~100 million years old) that is fascinating to observe. Three rays extend from Tycho: one passes between Mare Humorum and Mare Nubium, another heads south, and the last one heads towards Mare Nectaris. The moon's surface is filled with pleasant surprises. Figures 8.3 and 8.4 only show a few of them. Others are labeled in some of the images in chapter 12.

The line separating the lit and unlit parts of the moon is called the terminator. The terminator is a great target for amateur astronomers because so many craters are visible there (regardless of its location). When you use higher magnifications, you realize that the terminator is not the smooth line it appears to be with the unaided eye.

1	Mare Crisium
2	Mare Fecunditatis
3	Mare Nectaris
4	Mare Tranquilitatis
5	Mare Serenitatis
6	Mare Vaporum
7	Mare Imbrium
8	Oceanus Procellarum
9	Mare Insularum
10	Mare Nubium
11	Mare Humorum
12	Mare Frigoris
13	Mare Cognitum

Figure 8.3 – A more realistic map of the moon.

Observing the moon is incredibly entertaining and rewarding at *any magnification*. You can choose to use a moon filter or not, but a moon filter is something you should invest in (even if you are on a budget). A moon filter will lessen the intensity of the moon, enabling you to see more details on its surface. The terminator slowly moves across the moon as the day or night proceeds. For many beginner astronomers, the terminator is the most rewarding place to point their telescopes.

The moon always looks the same from Earth – we only see one side. The time it takes for the moon to orbit Earth is equal to the time it takes the moon to rotate once about its axis, so the motions cancel out. This is, however, a simplification of reality. In reality, we can actually see approximately 60% of the moon's surface. This is because the moon reveals a little more of its surface at extremes during its orbit around Earth.

Obviously, there are many other maria and craters that have names. Figure 8.4 identifies some of these. More moon maps can be found in chapter 12. The edge of the moon is called the limb. A cheap refractor telescope with a small f number (less than ~15) will make the limb appear to be colored.

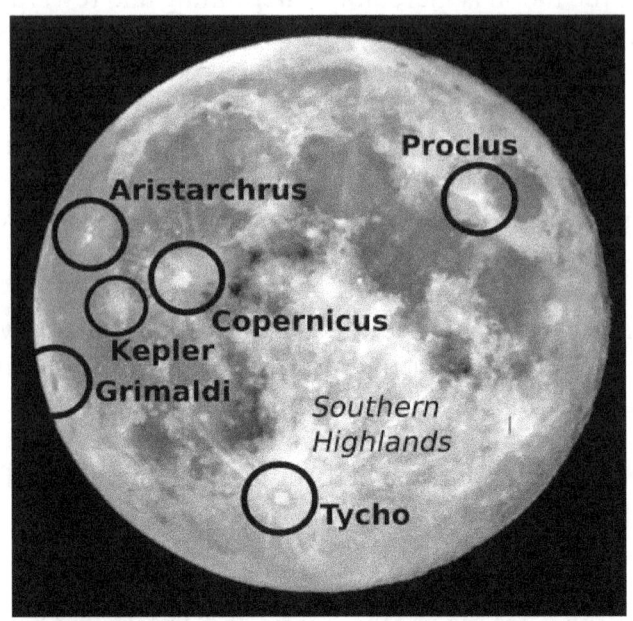

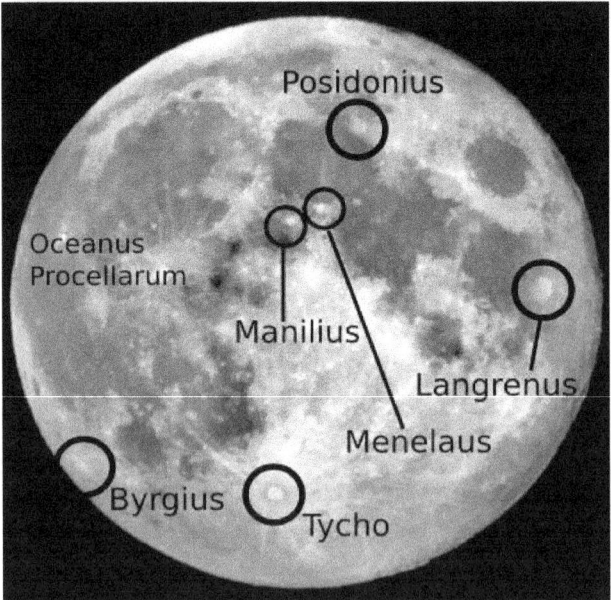

Figure 8.4 – Various craters on the moon.

The images in figure 8.4 were taken using a 50mm refractor telescope. Somewhat better images of the moon using a 90mm refractor can be found in chapter 12. Complete moon maps are readily available and cheap (if not free). Buy or download one if you enjoy finding the smaller craters on the moon's surface.

The phases of the moon are actually relatively difficult to visualize and understand. The moon completes one orbit around the Earth in ~27 days (as we have stated before). The moon moves relatively quickly through the night sky because the Earth rotates. The moon moves relatively slowly around the Earth, and this is why the terminator moves very slowly across the moon's surface as a night progresses.

Figure 8.5 shows that only the visible side of the moon faces the Earth at any given time. The "dark" side of the moon is not really dark at all, it gets just as much sunlight as the visible side of the moon over the course of the moon's orbit (we just cannot see it). At any point, like the point in figure 8.6, the moon is moving very slowly around the Earth, but the Earth is rotating relatively faster. This is difficult to visualize, but to a human on Earth, the moon in figure 8.5 would be a gibbous moon and the moon in figure 8.6 would be a quarter moon.

Figure 8.7 zooms in on the moon in an attempt to clarify any confusion. This is a difficult concept to grasp, but hopefully it will make more sense after you take some time to think about it. Figure 8.8 shows what phases occur at different points in the moon's orbit around the Earth. Figures 8.9 and 8.10 bring all of this information together.

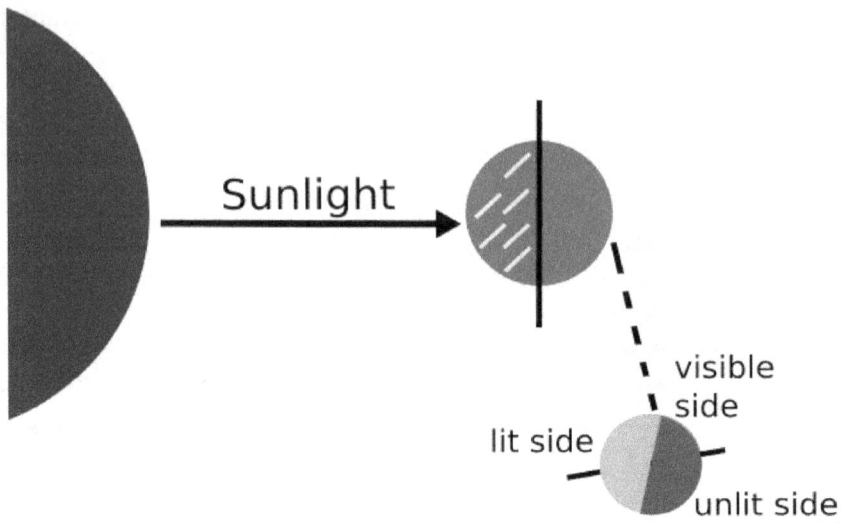

Figure 8.5 – Image to help the reader understand the moon's appearance at a point during its orbit around Earth. Roughly a gibbous moon.

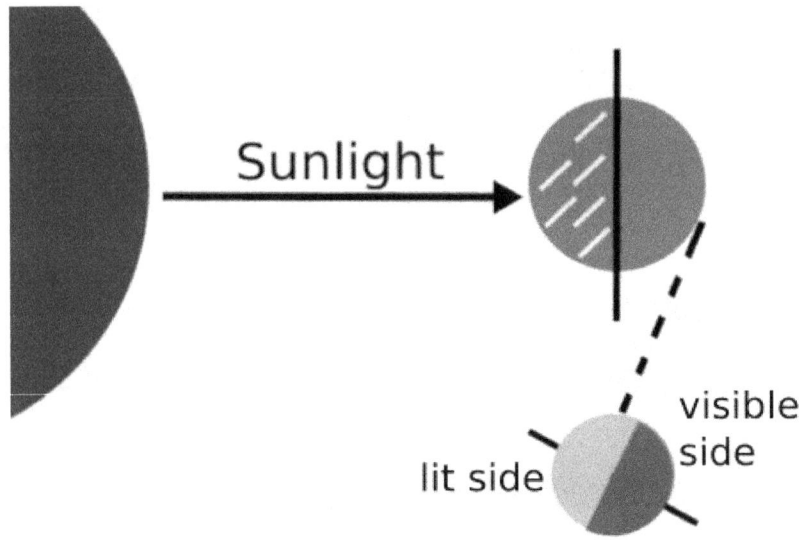

Figure 8.6 – Image to help the reader understand the moon's appearance at a point during its orbit around Earth. Roughly a first quarter moon.

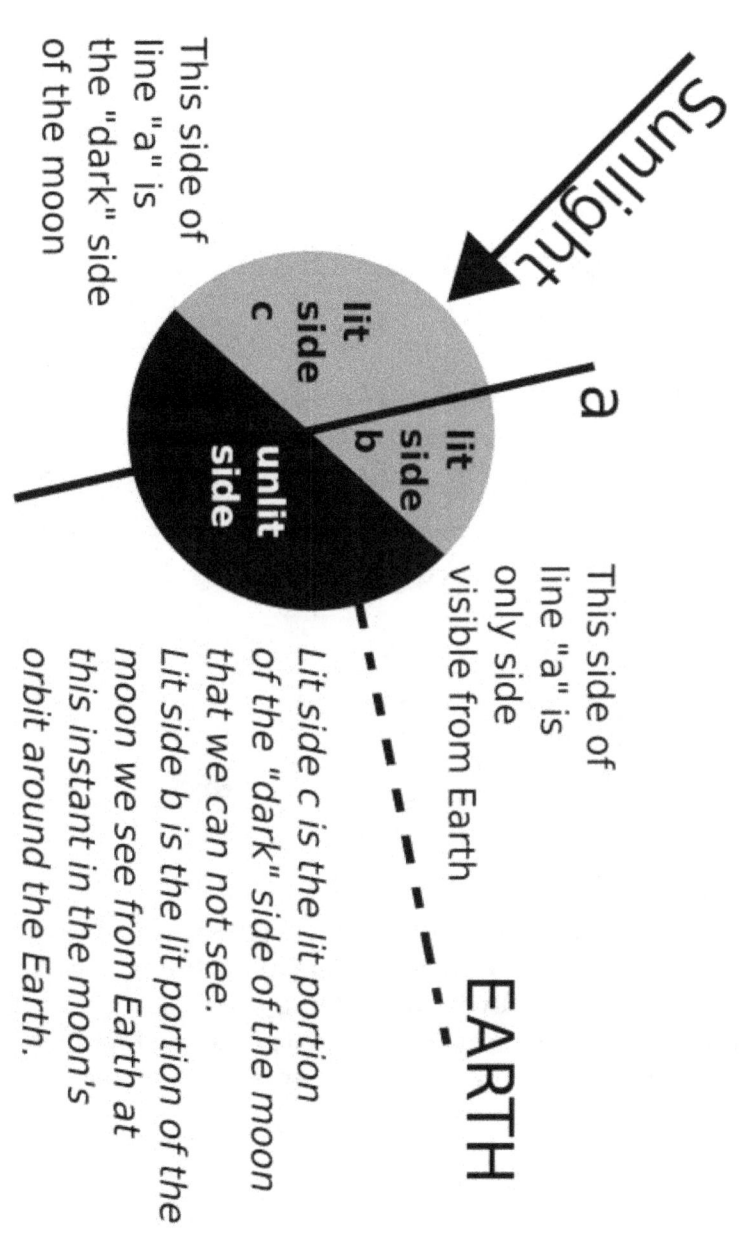

Figure 8.7 – Clarification of moon phase confusion. Roughly a
crescent moon.

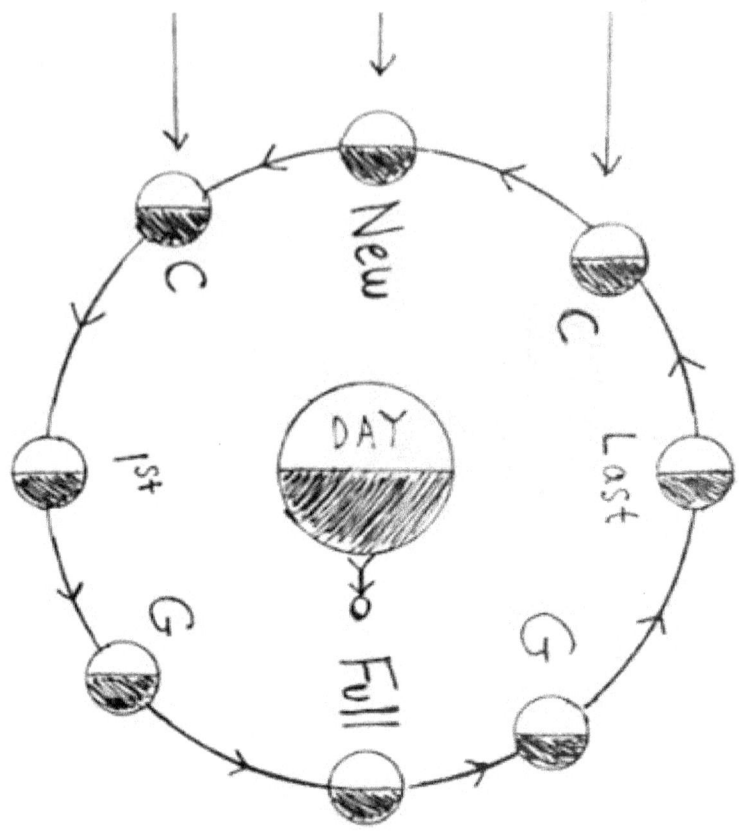

**Figure 8.8 – Phases of the moon (G = gibbous; C = crescent).
You (the stick figure) are observing at midnight. The Earth
rotates counterclockwise in this image. We can see that, at
midnight, the last quarter moon will rise and will travel
through the night sky as the Earth rotates.**

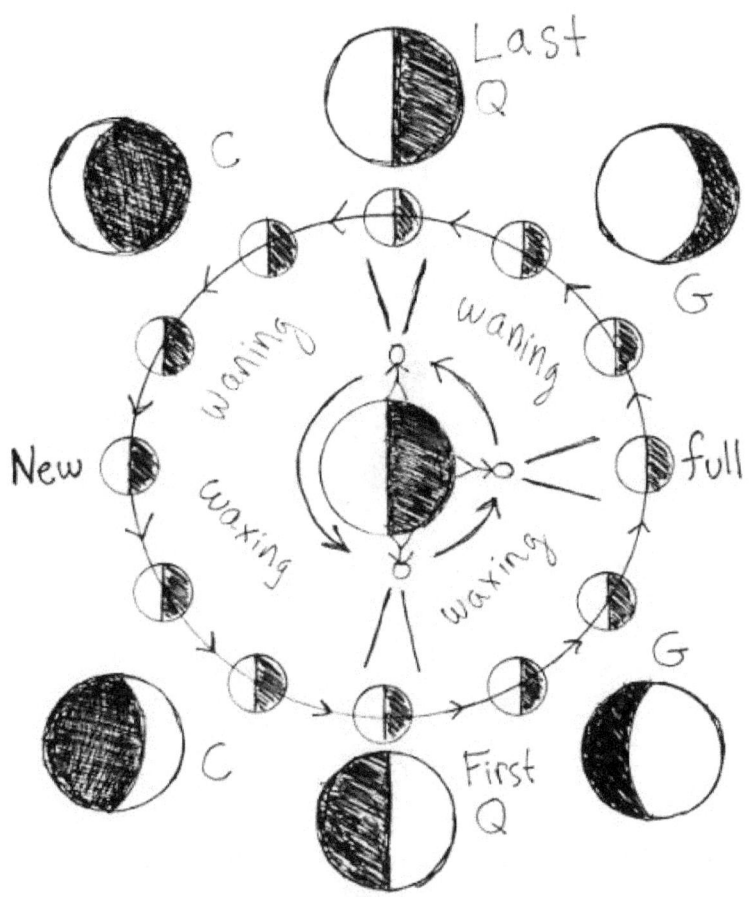

Figure 8.9 – (1) The west (left) side of the moon is lit during the waning moon. The east (right) side of the moon is lit during the waxing moon. (2) The first quarter moon is at its highest point in the sky at sunset. (3) The last quarter moon is at its highest point in the sky at sunrise. (4) The full moon sets around sunset and rises around sunrise. This is difficult to visualize. Imagine observing at sunset (upside down in figure 8.9). The full moon would just be rising through the horizon at this point. A similar analysis can be done for every phase.

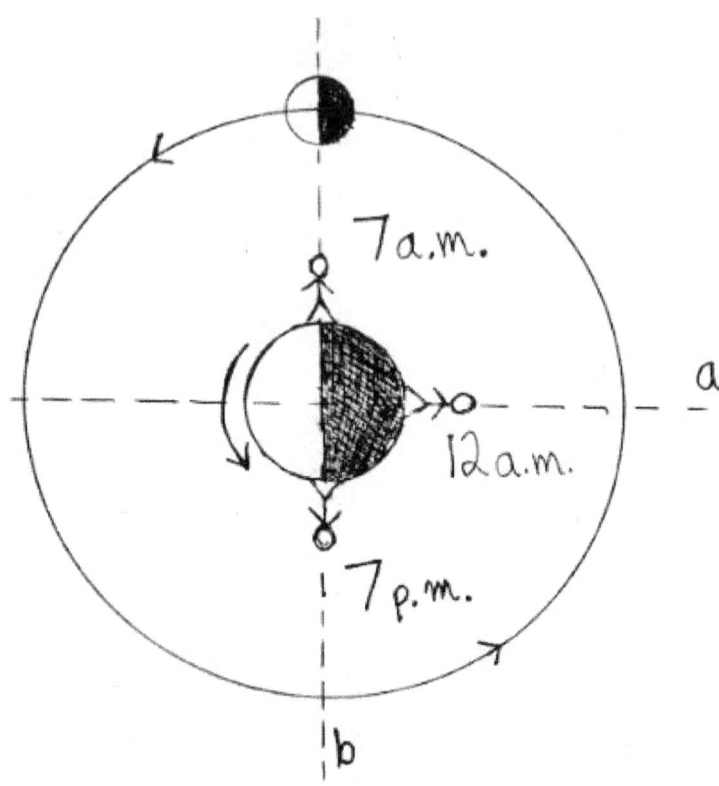

Figure 8.10 – Case study: the last quarter moon. (1) To the observer at 7pm, line *a* represents the horizon. The last quarter moon does not lie within the horizon boundaries, so the last quarter moon is not visible at 7pm. (2) To the observer at midnight (12am), line *b* represents the horizon. The last quarter moon can be found just rising through the horizon at midnight. (3) To the observer at 7am, line *a* represents the horizon again. At 7am, the last quarter moon is at its highest point in the sky. Try to perform your own case studies on the other phases of the moon.

Chapter 9

Other Observations

Constellations and Deep Sky Objects

Constellations are not all that great but they are useful because they make it easier to find some deep sky objects (DSOs). We are not all that interested in deep sky objects in this book because they are, generally, not all that great to look at with a cheap telescope (we will, however, discuss a few of the exceptions). As of 2017, there are many free programs and applications that make deep sky objects much easier to find. We will, however, talk about how to find these objects without those resources.

You do not have to use constellations to find DSOs, but it is one technique you can use. Unfortunately for you, I am the writer of this book and I prefer this method so we will use it (many people prefer the similar star-hopping method). I will introduce a series of constellations and then show you where to find some DSOs. I will show what the constellations look like, but not necessarily where to find them. Constellations are actually quite large. In fact, if you have never looked for constellations before, I think you will be surprised. You will be able to recognize constellations fairly easy after a few observation sessions. The rough location of some DSOs are given in the following drawings. You will need to look around to find them. The "hunt" for DSOs is part of what makes amateur astronomy such a fun hobby (if you are having trouble finding an object, try using a lower magnification first and then switching to higher magnifications).

The larger the aperture and the darker the skies, the better your chances are of finding these objects. Throughout this chapter, the objects in bold are the objects you should try first.

The Orion constellation is quite a popular one. In fact, you can probably already identify Orion's belt in the winter sky. Orion is only visible during the winter months, November to March, in the Northern Hemisphere. Orion can be seen in figure 9.1. This constellation is very easy to find.

The Orion Nebula, **M42**, is one of the few deep sky objects you should definitely try to observe with a cheap telescope. The Orion Nebula is quite large, so you may elect to use a lower magnification (~30x). Depending on where you are observing, you may only be able to see a small, fuzzy glow of light. There is a four-star system in the heart of the Orion Nebula called **the Trapezium**. You need to use a high magnification to see the Trapezium. The Orion Nebula is always listed as one of the objects every amateur astronomer should observe during the winter months. Orion contains several multiple star systems.

To find M42:
(1) Locate Orion.
(2) Locate the faint stars under the belt of Orion.
(3) Point your telescope in that direction with a low magnification and you will see M42.

Orion's belt points toward **Sirius**, the brightest star in the sky. Sirius, itself, is something you should look at with your telescope (it is actually a difficult double star). Under Sirius is an open cluster, **M41**, which is best visible from November to March. This can be seen in figure 9.2. Not much magnification is needed to see the M41 open cluster. M41 is similar in size to the full moon in the night sky. M41 can be found ~4 degrees (~8 full moons) south of Sirius. M41 is the first DSO I ever saw with a telescope.

M46 and **M47** are also visible this time of the year. They can be found approximately twice the distance to the left of Sirius that separates Mirzam and Sirius. They are both open clusters and are worth an attempt. **Betelgeuse** is a red super giant.

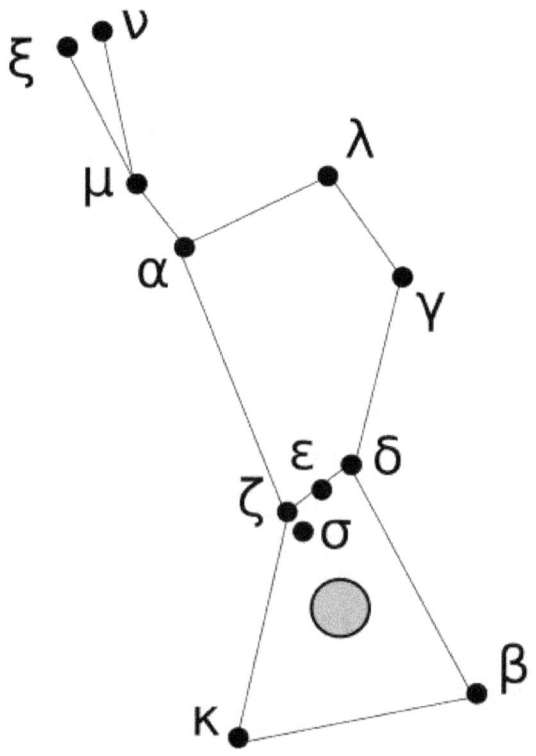

Figure 9.1 – The constellation, Orion. (Circle = M42)

Star in Figure 9.1	Common Name
α	Betelgeuse
β	Rigel
γ	Bellatrix
δ	Mintaka
ε	Alnilam
ζ	Alnitak
κ	Saiph
λ	Meissa

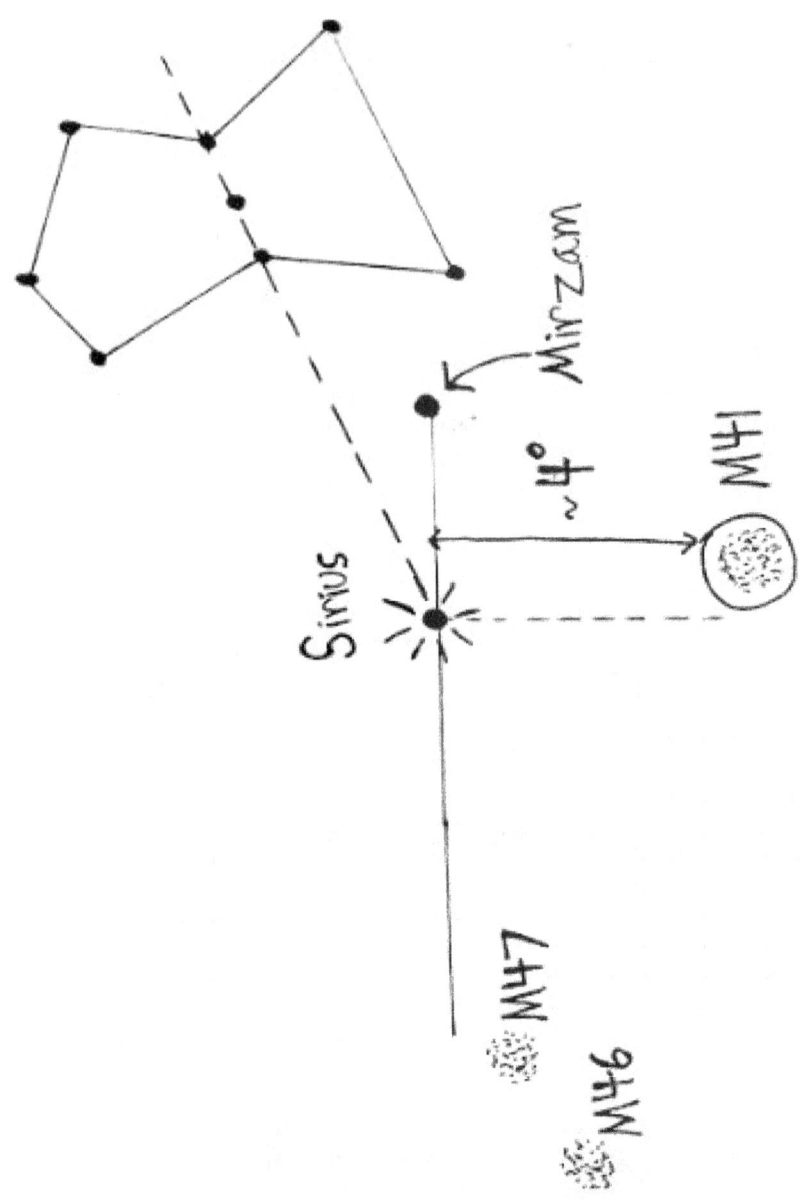

Figure 9.2 – Finding M41, M46, and M47 using Orion.

Sirius is part of the constellation, Canis Major. In fact, Sirius is the alpha star. Sirius can, therefore, be called alpha Canis Majoris. Mirzam is the beta star in Canis Major.

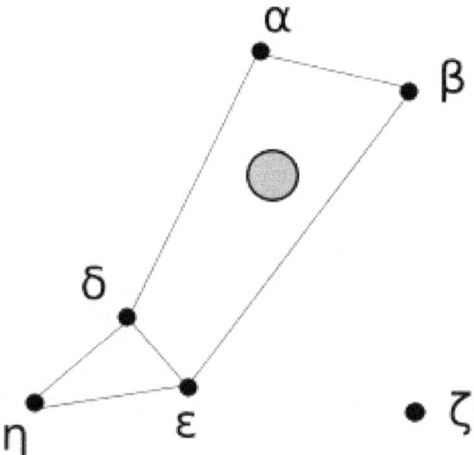

Figure 9.3 – Finding M41, M46, and M47 using Canis Major.

To find M41:
(1) Find Orion and Locate Orion's belt.
(2) Trace Orion's belt to Sirius (alpha Canis Majoris).
(3) M41 is located ~4 degrees south of Sirius.

To find M46 and M47:
(1) Find the complete Canis Major constellation, Sirius, and Mirzam (beta Canis Majoris).
(2) M46 and M47 are located to left of Sirius by twice the distance that separates Mirzam (beta Canis Majoris) and Sirius (alpha Canis Majoris).

If you are curious about what these objects look like, refer to the images in chapter 11 or search for them elsewhere.

Orion has a neighbor constellation, Taurus. **Aldebaran** is the alpha star in this constellation and it is a huge red giant star. The diameter of Aldebaran is ~35 times larger than the diameter of the sun. **M45** is known as the Pleiades and it is a bright star cluster (you area able to see it with an unaided eye). M1, the crab nebula, is a challenge for small telescopes. Taurus is best visible from November to March.

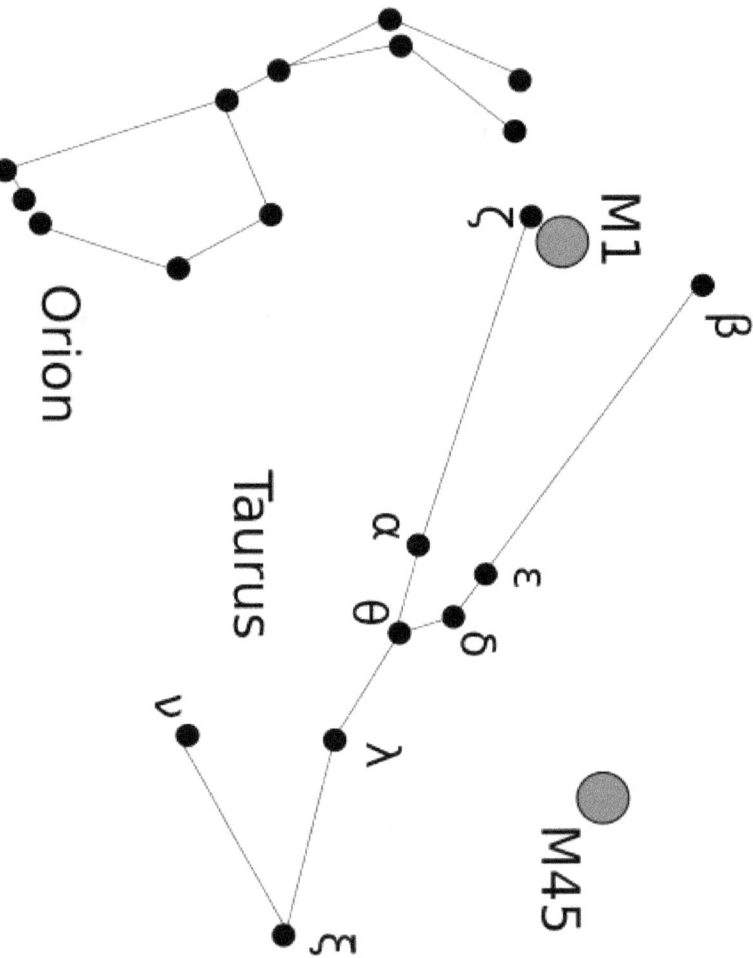

Figure 9.4 – Taurus Constellation.

Ursa Major and Ursa Minor, the "big dipper" and the "little dipper", respectively, are circumpolar constellations. This means that they are always visible in the Northern Hemisphere (Ursa Minor – see figure 9.5; Ursa Major – see figure 9.6). Cepheus, Cassiopeia, and Draco are also circumpolar constellations.

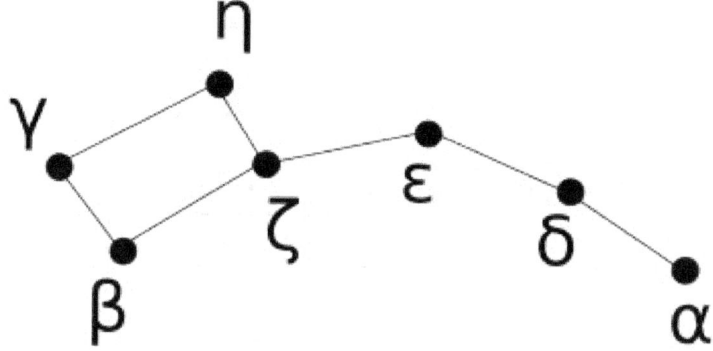

Figure 9.5 – The "little dipper", Ursa Minor.

Star in Figure 9.5	Name
α	Polaris
β	Kochab
γ	Pherkad
δ	Yildun
ε	Urodelus
ζ	Ahfa
η	Anwar

An arrow starting at Merak and extending through Dubhe (in Ursa Major – next page) will run into Polaris. **Polaris** is called the "North Star" because it is located very close to the northern pole. Polaris is a double star, but one star overpowers the other, making it difficult to resolve.

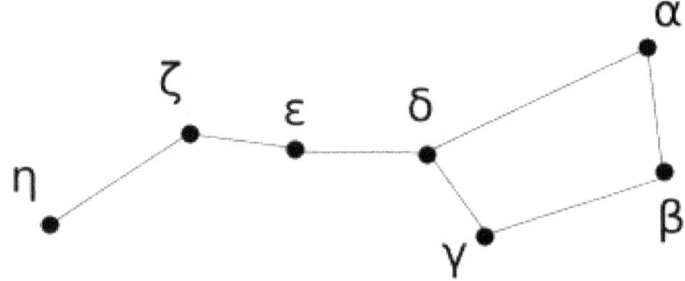

Figure 9.6 – The "big dipper", Ursa Major.

Star in Figure 9.6	Name
α	Dubhe
β	Merak
γ	Phecda
δ	Megrez
ε	Alioth
ζ	Mizar
η	Alkaid

Near **Mizar** is the star Alcor, and they appear to be a single star to someone with poor eyesight. A telescope can easily split these two if your eyes cannot. Mizar is, itself, a double star and a telescope can split it. You can, therefore, consider the Alcor-Mizar system a triple star if you have poor eyesight. If you happen to see a faint star between these two stars, it is Sidus Ludoviciana (8th magnitude). In the 18th century, Johann Liebknecht announced that it was a new planet (he was wrong).

Many galaxies and nebula surround Ursa Major. **M51**, the whirlpool galaxy, is located just under Alkaid in figure 9.6. M51 is a difficult object to see with an aperture less than 8 inches.

Cepheus is circumpolar but is, unfortunately, fairly boring.

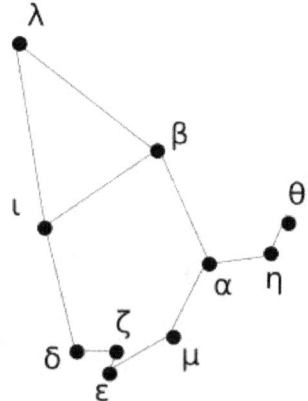

Figure 9.7 – Cepheus Constellation.

Draco is also circumpolar and is also, unfortunately, fairly boring.

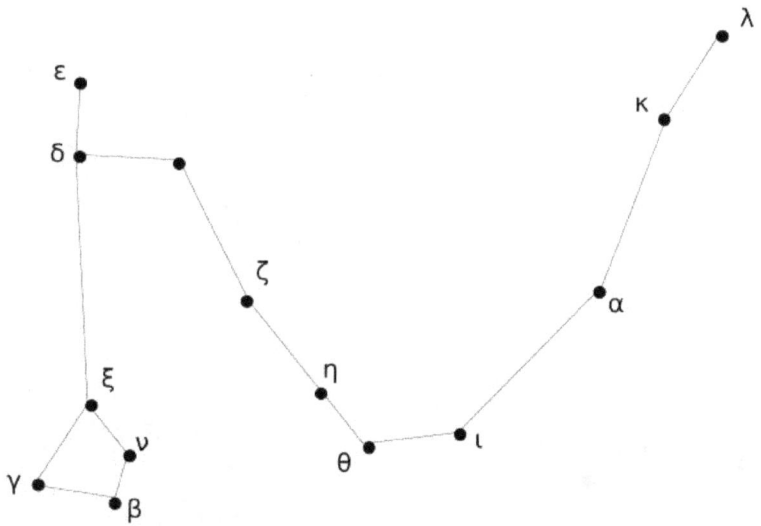

Figure 9.8 – Draco Constellation.

Both Cepheus and Draco contain some multiple star systems. They are included in table 9.2.

Cassiopeia is a popular constellation as well. It is always visible, but is best visible during the fall. **Eta Cassiopeia** is a double star.

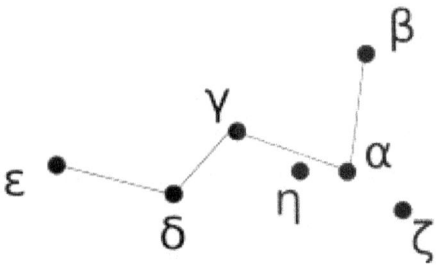

Figure 9.9 – Cassiopeia Constellation.

A double cluster containing **NGC 869** and **NGC 884** can be found between Cassiopeia and Perseus. Look between the gamma star in Cassiopeia and the eta star in Perseus. They are best observed from October to February. We will discuss M31 later in this chapter.

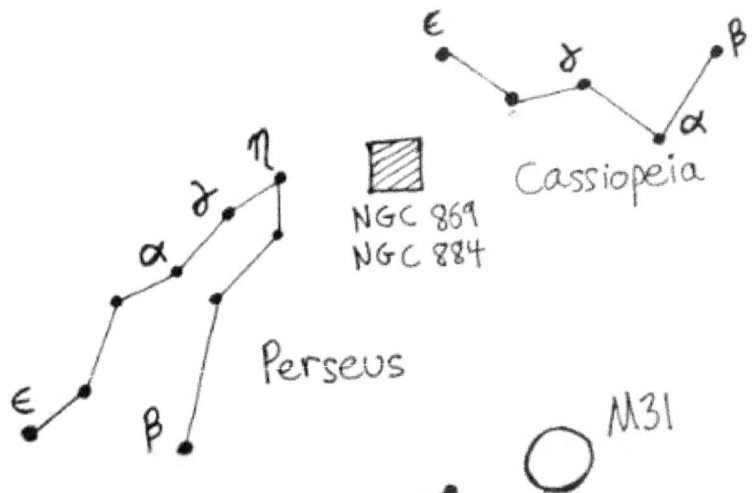

Figure 9.10 – Finding the NGC 869/884 double cluster.

In the early summer, Leo is a recognizable constellation. The same line that connected Dubhe and Merak in Ursa Major and led to Polaris runs into Leo as well, but on the other side of Ursa Major (see figure 9.12). Leo is shown below. Leo is best visible from February to June in the Northern Hemisphere.

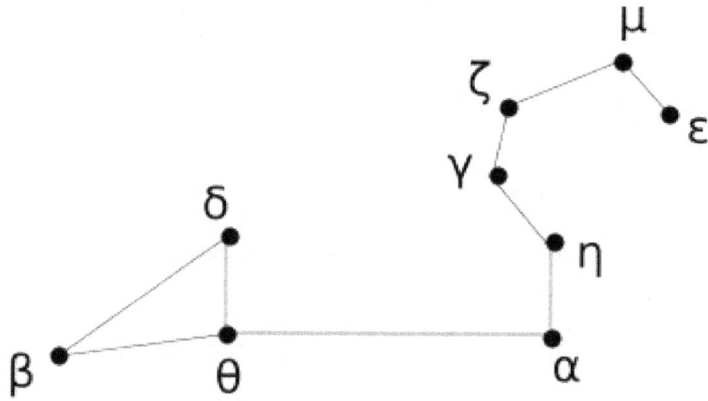

Figure 9.11 – The constellation, Leo.

Star in Figure 9.11	Common Name
α	Regulus
β	Denebola
γ	Algieba
δ	Zosma
ε	Algenubi
ζ	Adhafera
θ	Chertan or Chort

The backwards "?" is often referred to as the "sickle" and it is easily identifiable in the night sky. **Algieba** is a double star that can be split with a magnification of ~100 times. It is a difficult star to split but it is worth the effort. The "Leo Triplet" is a group of three galaxies that resides under Chertan in figure 9.11. It consists of **M65**, **M66**, and **NGC 3628** in a triangular shape, but they are very difficult to see. You need dark skies and a large aperture. Try it if you are up for a challenge (I was actually hesitant to even include the Leo Triplet in this book).

To find the Leo Triplet:
(1) Identify Leo and locate Chertan.
(2) You will see a star southwest of Chertan. Imagine a line that connects these two.
(3) The Leo Triplet can be found close to that line.

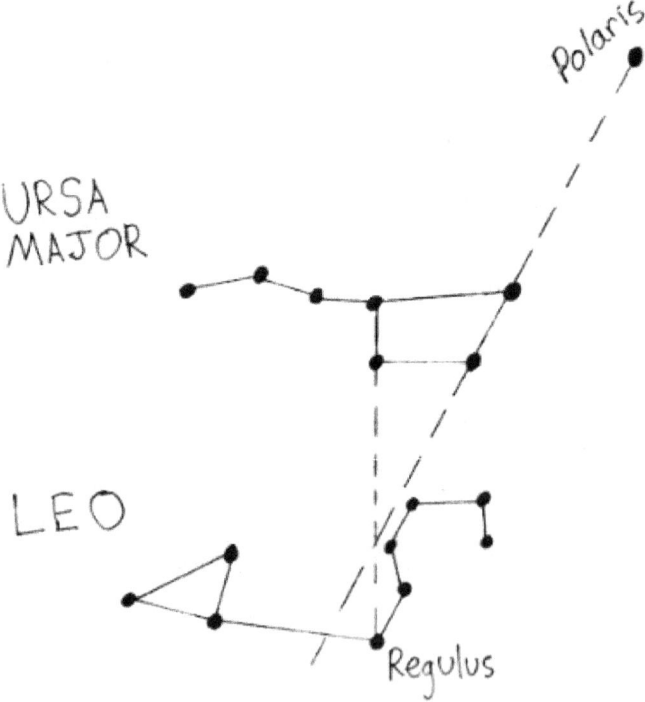

Figure 9.12 – Intersection of Leo, Polaris, and Ursa Major.

Cancer is another constellation that has some interesting objects around it. Cancer is shown in figure 9.13. Cancer is seen in the Northern Hemisphere from January to May. Cancer is a faint constellation, but it is close to Leo and the sickle faces it (see figure 9.14). Cancer's brightest star has a magnitude of ~3.5 so the entire constellation may be invisible in light-polluted skies. Cancer is a victim of light pollution and this makes finding M44 a difficult but worthwhile task.

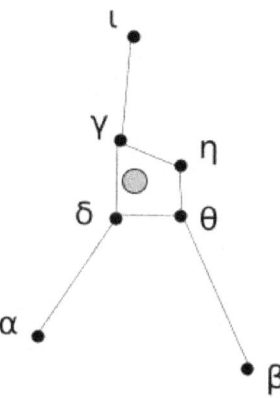

Figure 9.13 – The constellation, Cancer. (Circle = M44)

Star in Figure 9.13	Common Name
α	Acubens
β	Altarf
γ	Asellus Borealis
δ	Asellus Australis
ι	Iota Cancri

The Beehive cluster, **M44**, is massive group of stars located between Asellus Australis and Asellus Borealis and slightly to the right. M44 is a faint, large object and it should be visible at low magnifications. Figure 9.14 summarizes some of what we have discussed thus far.

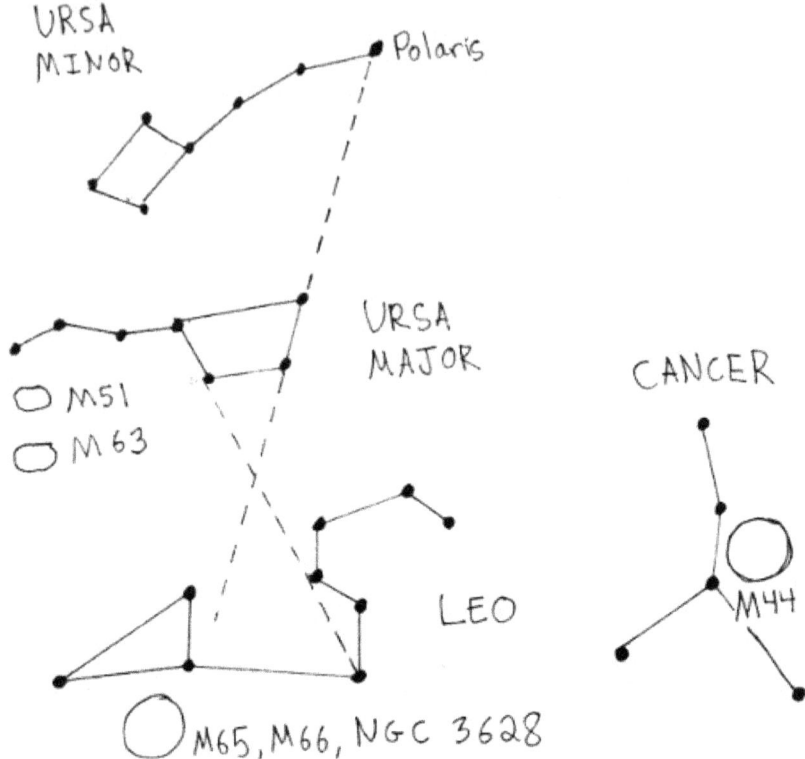

Figure 9.14 – Poorly drawn sky map showing most of what we have discussed so far.

Since I have some room on this page, I want to mention two things: (1) You are free to observe whatever you want with your telescope. You want to look at Regulus? Sure! Go for it. Do whatever you want. I am simply listing some of the more interesting and easy-to-find objects. (2) I am listing the approximate months these objects can *best* be observed during. You might be able to observe these objects early in the night or early in the morning, before or after these given time periods. Just go outside and see what constellations are in the sky. There will always be something in the night sky you can point your telescope towards.

From May to October, the skies are quite different. The star, **Vega** (alpha Lyrae), is a bright blue star with a magnitude of ~0.3. The summer triangle is made up of Vega (part of Lyra), Deneb (part of Cygnus), and Altair (part of Aquila).

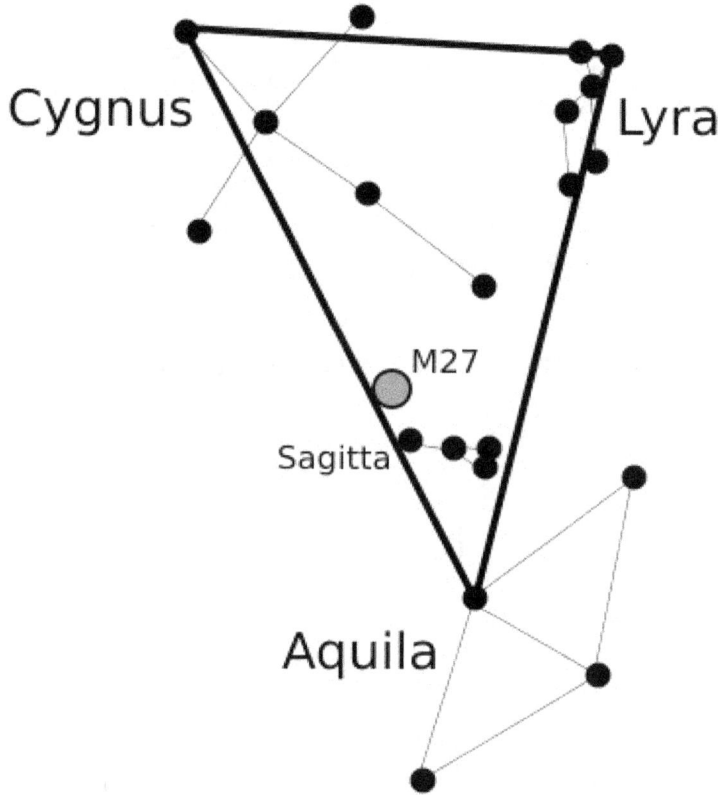

Figure 9.15 – The summer triangle.

M27, the "dumbbell" planetary nebula, is located between Deneb and Altair, but closer to Altair. It is spectacular to look at, but you need medium magnification and dark skies. M27 is best seen from June to October. **M57**, the ring nebula, can be found between the two southern-most stars in Lyra. Magnifications of ~100 times can generally resolve the basic ring shape.

Lyra and Cygnus are popular summer constellations. **Epsilon Lyrae** is a multiple star system (it is actually a double-double).

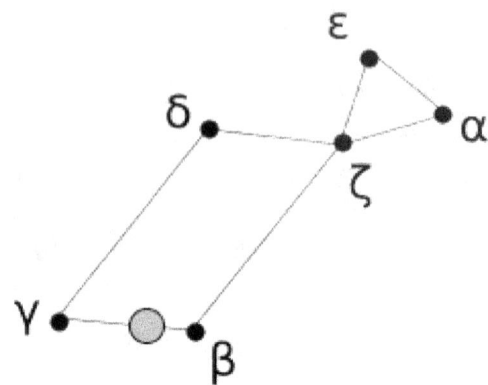

Figure 9.16 – Lyra Constellation. (Circle = M57)

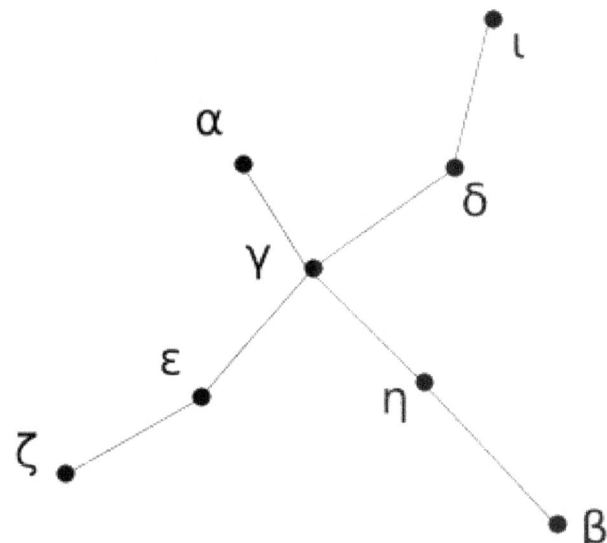

Figure 9.17 – Cygnus Constellation.

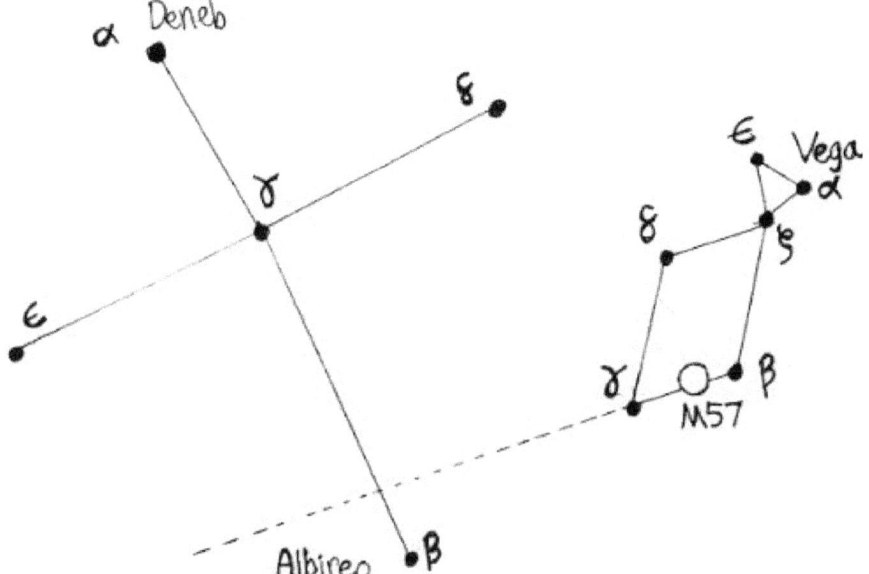

Figure 9.18 – The constellations Lyra and Cygnus.

Cygnus contains the "Northern Cross". Alpha Cygni (Deneb) and Beta Cygni (Albireo) are two very important stars. The former is important because it is part of the summer triangle (as previously stated). **Albireo** is important because it is a fantastic double star. The two stars in this double star system are drastically different colors. This makes Albireo a valuable target for amateur astronomers. Albireo can be split with magnifications as low as 30 times. **M56** is a globular cluster that can be found between gamma Lyrae and Albireo.

To find M57:
(1) Find Vega (it is a bright, blue star).
(2) Find the entire Lyra constellation.
(3) Draw a mental line that connects gamma and beta Lyrae.
(4) M57 lies on this line, but closer to beta Lyrae.

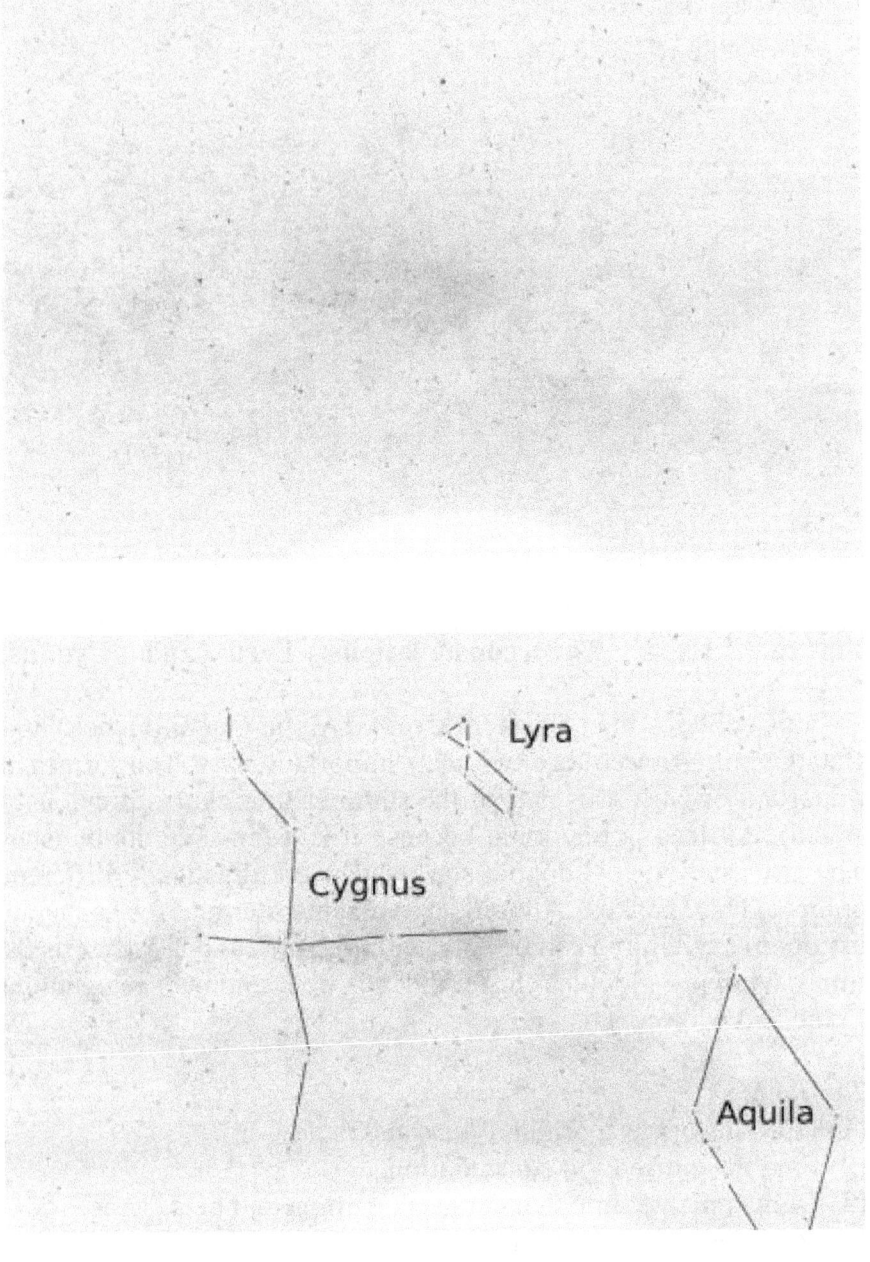

If you can find the summer triangle, you can find many other objects in the summer sky. If you can find Vega, you should be able to find Hercules; and if you can find Hercules, you should be able to find **M13** and **M92**. M13 and M92 are best observed from May to October. Both require medium magnifications, but M13 can be seen in almost any sky conditions. **Alpha Herculis** is a multiple star system (the brightest star is called Rasalgethi).

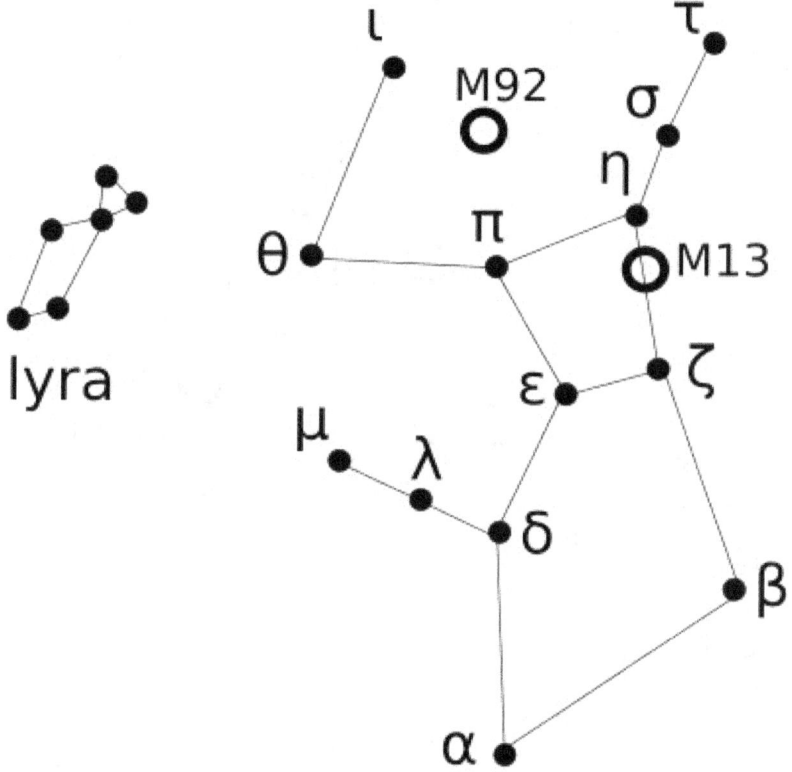

Figure 9.19 – Hercules, M13, M92, and alpha Herculis.

Talk to any amateur astronomer and you will most likely find that their favorite globular cluster is M13 (or M3). It is a very popular globular cluster because it is bright and relatively easy to find. It is definitely worth a view through a small telescope. M92 is also a great globular cluster, but M13 is more popular.

If you can find "the teapot" you can find many great objects worthy of observation. "The Teapot" can be found in the constellation, Sagittarius, and can best be observed from August to September. Find this, and you have a lot to observe. The Lagoon Nebula, **M8**; The Trifid Nebula, **M20**; A globular cluster, **M28**; A cluster, **M22**; the butterfly cluster, **M6**; and Ptolemy's cluster (an open cluster), **M7**, are all visible during this time (see figure 9.20). **M21** is an open cluster that can be found above M20. You will find many other nebulae and clusters in this area in addition to the ones listed here.

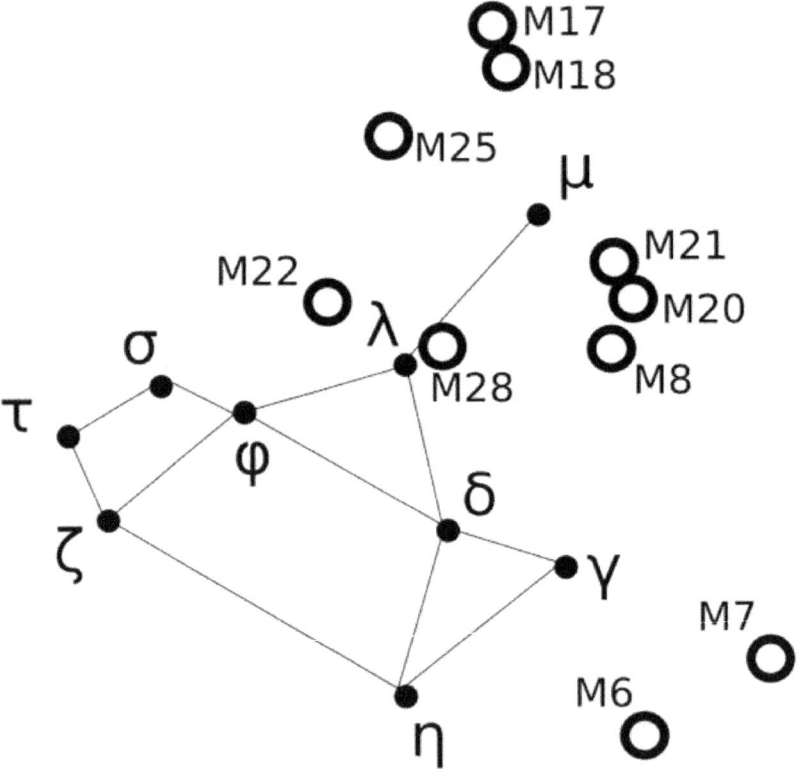

Figure 9.20 – The Teapot.

Next to the teapot is the constellation, Scorpius. **Antares**, Alpha Scorpii, is a red super giant that is clearly red, even to the unaided eye. **M4** is another faint globular cluster. It lies close to Alpha Scorpii and Sigma Scorpii, so it is relatively easy to find.

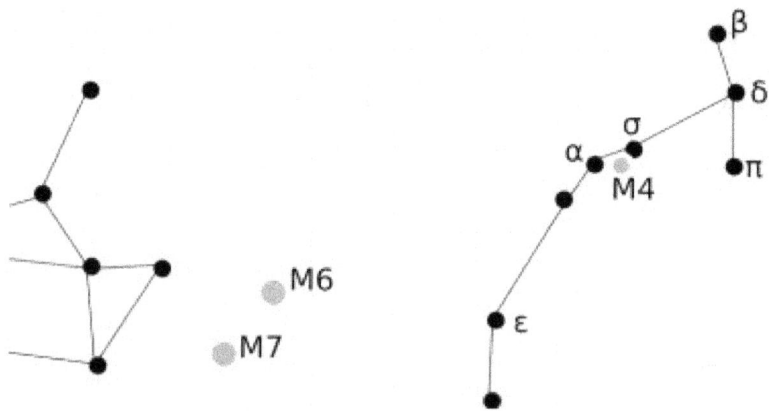

In the following image, M4 is visible. The following image is only 25 stacked 1.6 second exposures, so it had to be edited to bring M4 out. This explains why the image is noisy. Regardless, it shows M4's location.

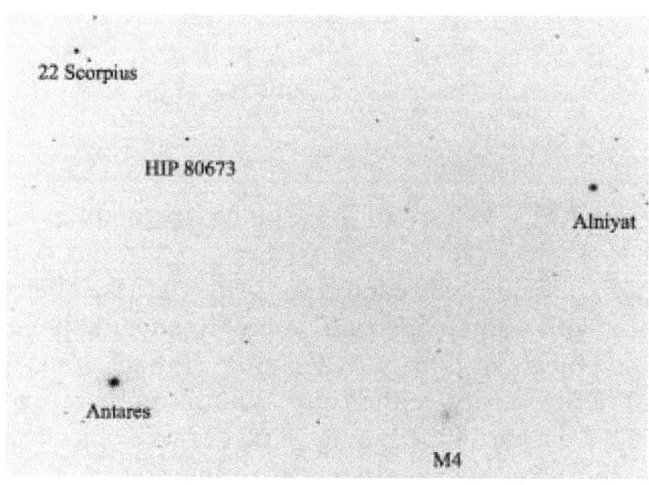

The Andromeda galaxy, **M31**, can be found near the "great square" in the constellation Pegasus (as shown in figure 9.21). It is best seen from September to January. The Andromeda galaxy is the closest major spiral galaxy to our Milky Way galaxy. It is 2.3 million light-years away. It takes light 2.3 million years to get to us from the Andromeda galaxy. You can see it with your eye alone (of course, in very dark skies). You will want to use the lowest magnification you can. M31 is huge.

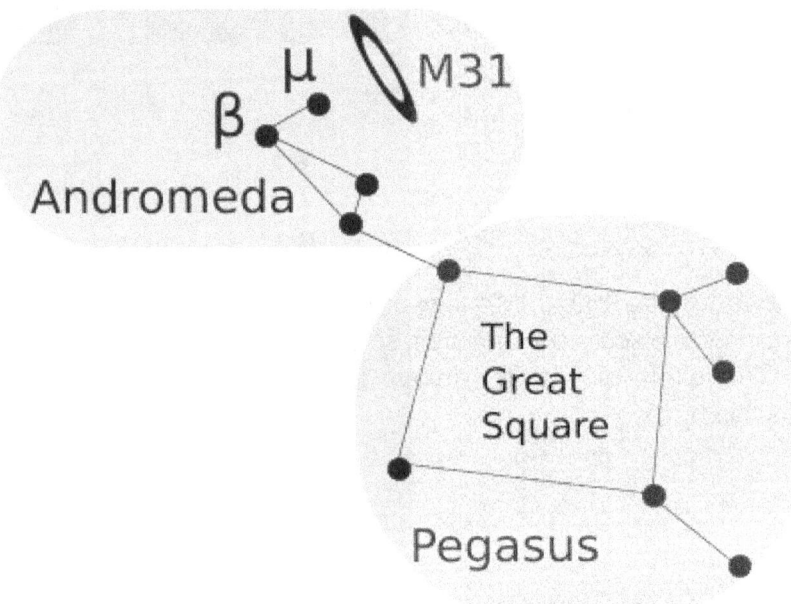

Figure 9.21 – Where to find the Andromeda galaxy.

Do not get frustrated if you cannot see some of these objects. If you are working with a small aperture, you will most likely not be able to see these objects in light-polluted skies. In fact, I can only take images of DSOs using either my 90mm refractor or 130mm reflector. I have a hard time seeing some of these objects at all with my 50mm refractor, but I do live near a major city.

Object	~Best Time	Magnification
M42	Nov. - March	Low – Dark Skies
Trapezium	Nov. - March	High
Sirius	Nov. - March	Any
M41/46/47/45	Nov. - March	Low
Mizar-Alcor	Always	Any
Leo Trio	Feb. - June	Med. – Dark skies
Algieba	Feb. - June	Any
M44	Feb. - April	Low
M27/57	June - Oct.	Med. – Dark Skies
Vega/Albireo	June – Oct.	Any
M13/92	May - Oct.	Medium
M6/7/8	August – Sept.	Med. – Dark Skies
M20/21/28	August – Sept.	Med. - Dark Skies

Table 9.1 – Summary of recommended observations.

You will notice that, in table 9.1, none of the nebulae or clusters require high magnification. Magnification is, in general, not the name of the game when it comes to nebula and cluster observation (since these objects are fairly large). Aperture and light pollution are the most important factors. This does not necessarily mean that you cannot see these objects with a cheap 50mm telescope, but it will be more difficult if you live in a light-polluted area. LPR filters can sometimes help, but they will usually only help the observation if you are using a large aperture.

You may or may not be wondering why the night skies look different in June than they do in December. This is actually relatively easy to visualize – see figure 9.22.

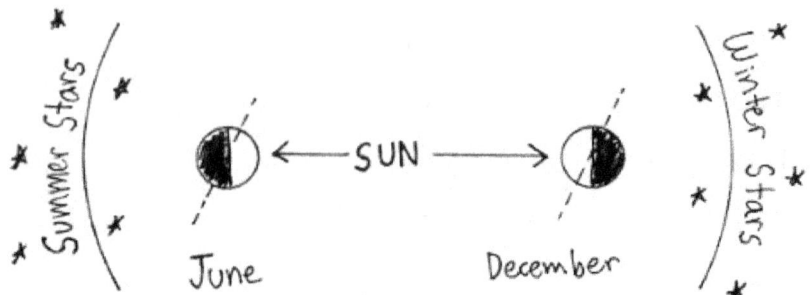

Figure 9.22 – The stars in the summer night sky are not the same as the stars in the winter night sky.

In figure 9.22, during June, an observer in the Northern Hemisphere can only see the stars, constellations, and DSOs to the left. During December, an observer in the Northern Hemisphere can only see the stars, constellations, and DSOs to the right. Obviously the "left" stars and the "right" stars are different, so only certain stars, constellations, and DSOs are visible during certain parts of the year. The exceptions are circumpolar constellations, which are always visible. The northern pole points to objects like Ursa Minor, and this is why Ursa Minor is always visible in the Northern hemisphere.

The nebulae, clusters, and stars I have listed in this chapter are just a few of the objects you can observe. There are many resources that list objects to look for with a small telescope, but the ones listed here are generally the easiest ones to find and the most rewarding to look at once you find them. Find these first and then move onto more difficult objects.

Just like stars, DSOs have an associated magnitude that tells us how bright they are in the night sky. The lower the magnitude, the brighter it is. Distance from the Earth is also included to show that the two are not necessarily related.

Messier	Magnitude	Distance (light-years)
M42	4.0	1600
M41	4.6	2300
M44	3.7	577
M46	6.0	5400
M47	5.2	1600
M51	8.4	37 million
M63	8.6	37 million
M97	9.9	2600
M27	7.4	1250
M28	6.8	18600
M8	6.0	5200
M6	4.2	1600
M7	3.3	800
M57	8.8	2300
M20	9.0	5200
M31	3.4	3 million
M13	5.8	25100
M35	5.3	2800
M45	1.6	380
M56	8.3	32900
M101	7.9	27 million

Resolving and Splitting Multiple Stars

What exactly defines a "split" star? Depending on how much magnification you use, a multiple star system can appear to be (1) a single star still, (2) a stretched star, (3) multiple resolved stars, or (4) several, split stars.

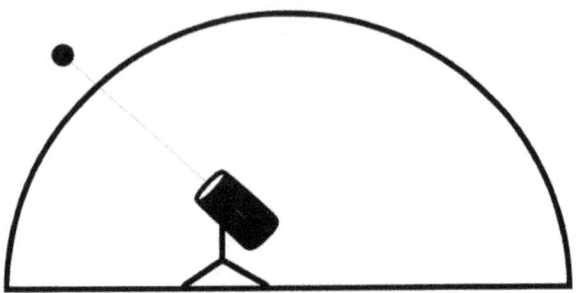

Figure 9.23 – A multiple star appears as a single star from Earth. The dome is the Earth's atmosphere.

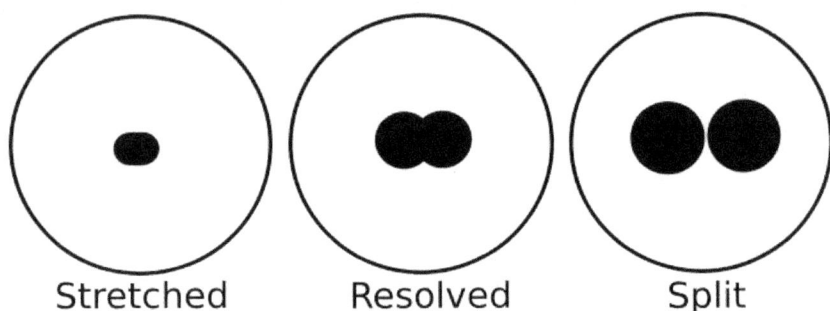

Stretched   Resolved   Split

Figure 9.24 – The star system can be "stretched" (lowest mag.), "resolved", or "split" (highest mag.)

I do not want to go into too much detail, but a table on the next page lists some of the popular multiple star systems to start with. If you find that you enjoy splitting stars, there are many resources that list multiple star systems you can use to find more. This is a good list to start with. Some are challenging, so you may not be able to resolve or split them.

Star System	Separation	Magnitudes
Mizar (+Alcor)	14.4", (709" M-A)	2.3, 4.0, 4.0
Polaris	18.4"	2.0, 9.0
Eta Cassiopeia	12"	3.4, 7.5
Beta Orionis	9.5"	0.1, 6.8
Delta Orionis	52.6"	2.2, 6.3
Iota Orionis	11.3"	2.8, 6.9
Lambda Orionis	4.4"	3.6, 5.5
Sigma Orionis	12.9", 43" (mult.)	4.0, 7.5, 6.5
Epsilon Canis Majoris	7.5"	1.5, 7.4
Alpha Leonis	177"	1.4, 7.7
Gamma Leonis	4.4"	2.2, 3.5
Epsilon Lyrae	208", 2.6", 2.3" (mult.)	All 5-6 range
Zeta Lyrae	44"	4.3, 5.9
Beta Lyrae	46"	3.4, 8.6
Beta Cygni	34.4"	3.1, 5.1
Beta Cephei	13.3"	3.2, 7.9
Delta Cephei	41"	3.9, 6.3
Nu Draco	62"	4.9, 4.9

Table 9.2 – Multiple star systems for small telescopes.

Figure 9.25 – Mizar and Alcor are clearly split, but Mizar is resolved since there is no clear gap between the two stars.

Stars can be difficult to resolve for many different reasons. Perhaps the simplest reasons to understand are: (1) how close the stars are to each other and (2) how bright the primary star is. If one star in the star system is significantly brighter than the other stars in the same system, the brighter star will over-power the dimmer stars. In this case, you can simply use higher magnifications. Sirius, for example, is a double star but since Sirius is so bright, it is difficult to resolve.

There is no set "best magnification" for splitting or resolving double stars. It depends on how close they are and how bright one star is with respect to the other(s). Some star systems can be split with magnifications as low as 30 times.

Take the Orion multiple star systems for example.

Beta Orionis: a challenge for smaller telescopes because one star is substantially brighter than the other. You should, therefore, use the highest magnification you can.

Delta Orionis: generally resolvable.

Iota Orionis: generally resolvable.

Lambda Orionis: use anywhere from 70 times to 100 times magnification.

Sigma Orionis: a large triple star system. Use lower magnifications.

Theta 1 Orionis: "The Trapezium". *Separations*: 8.8", 13", and 21.5". *Magnitudes*: 6.7, 7.9, 5.1, 6.7. When Orion is up, you should try to split this star system. The view is well worth the effort.

Epsilon Lyrae (bottom left) and Vega (top right). 23x.

Epsilon Lyrae. 130x.

Mizar (top left) and Alcor (bottom right). 23x.

Albireo (Beta Cygni). 70x. The best double!

Shooting Stars and Meteor Showers

A shooting star is really just a chunk of space dirt or space rock, called a meteoroid, falling through and burning up in the Earth's atmosphere. The burning meteoroid produces a streak of light in the night sky, called a meteor. People often call these streaks "shooting stars", but really they are just meteors. Any remaining material that survives the atmosphere and lands on the surface is called a meteorite.

Certain meteor showers occur during certain points in Earth's orbit around the sun (certain times in a year) and seem to originate from certain constellations. They do not actually come from these constellations, it just appears that way from Earth. The "peak time" is the time where you can see the most amount of meteors per hour. During these points in the year, you can see many "shooting stars".

Meteor Shower	Constellation	Approx. Time of Year
Quandrantids	Bootes	Late Dec. - Mid Jan.
Lyrids	Lyra	April 16 – April 27
Perseids	Perseus	Mid August
Draconids	Draco	Early-Mid October
Orionids	Orion	Mid-Late October
Leonids	Leo	Mid November
Geminids	Geimini	Early-Mid December
Ursids	Ursa Minor	Mid-Late December

Moissanite (Silicon Carbide), a great diamond alternative, is an incredibly rare stone that can be found in meteorite. Of course, the moissanite sold to the public is made in a lab.

Planets

Deep sky objects like nebula and clusters are usually worth the trouble it takes to find them. That being said, planets are still my favorite targets to observe with a telescope. I think (and you probably do too) that the planets in our solar system are incredibly interesting objects. Observing the rings around Saturn or the cloud bands of Jupiter brings out an amateur astronomer's inner-child. Magnification is most definitely the name of the game in planetary observation. You will be able to see planets with any telescope, but the detail you get is dependent on the magnification and the aperture of your telescope.

Venus

Mercury and Venus go through phases like the moon because they are located between the Earth and the sun. When Venus is up it is extraordinarily bright – it is the brightest object in the sky. Venus is never up much longer after the sun sets so observe Venus early in the night before it goes away. Venus has a thick and cloudy atmosphere, so you cannot see any surface detail. The phases of Venus, however, make the planet worthy of observation. I usually use a blue filter while observing Venus.

Mars

Mars is brightest in the sky when it is at opposition (see next page). Mars will look like a bright red star. Mars has ice caps at its poles that are better visible with filters. Mars also has maria and craters, just like the moon. Mars is difficult to observe with detail, especially with a small telescope.

Opposition

When a planet is at opposition, it is a big deal. This means that it is at its closest point to the Earth during its orbit around the sun. It will therefore, be bigger, brighter, and easier to observe with detail at that point. There are many resources that will tell you when planets will be at opposition, but go ahead and observe them any time you can and want to.

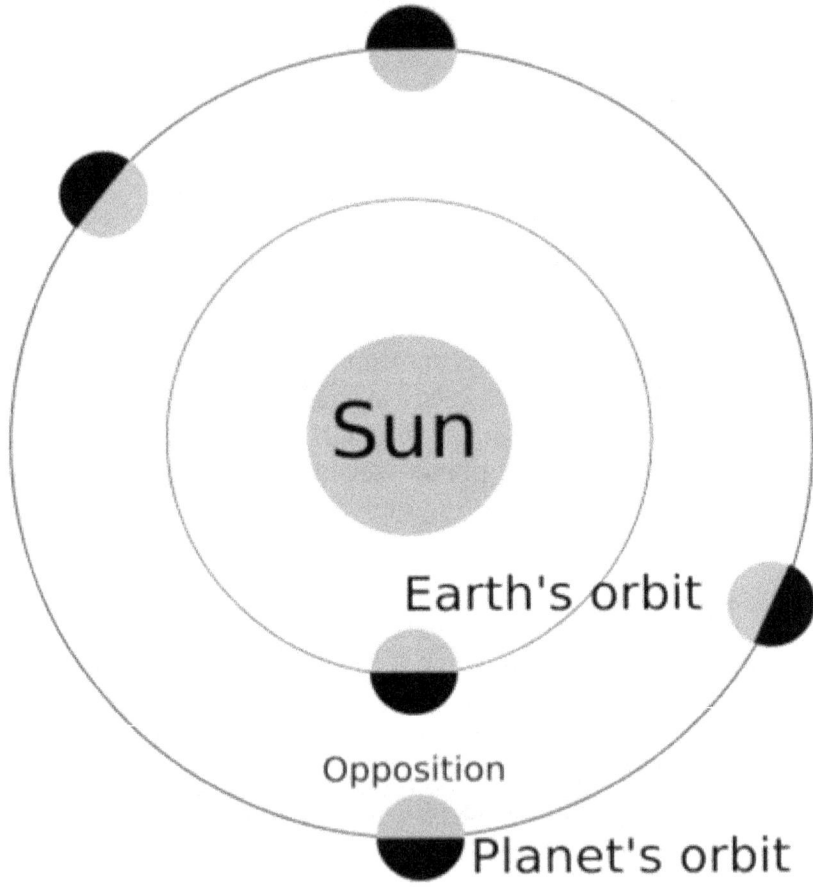

Figure 9.26 – When a planet is at opposition it is close to Earth and easier to observe with a telescope.

Jupiter

Jupiter is an absolutely fascinating object to observe if you can find it and if it is in the sky at the time you are observing. Jupiter may be in the west, east, or not even visible depending on what year and what time of year you are observing in. There is no easy pattern behind this, you just have to have a resource that tells you if Jupiter is visible at that time and where to look.

When you observe Jupiter at low magnifications and with a small telescope, you will not get anything special (but the view is still worth it). Figure 42 shows two images of Jupiter, one with the visible moons labeled. I used my 50mm refractor with a 9mm eyepiece to obtain these images (66.667 times magnification).

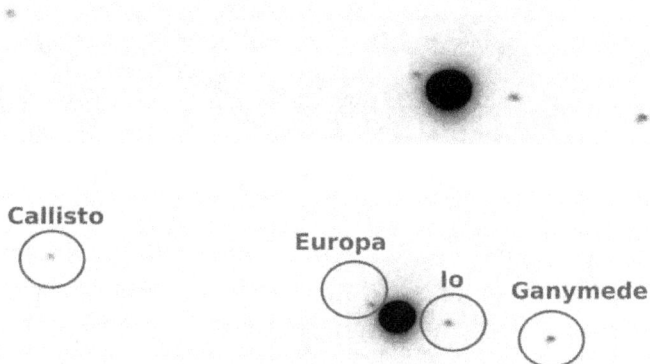

Figure 9.27 – Jupiter at ~70x magnification.

You will see four of the largest moons of Jupiter: Callisto, Europa, Io, and Gaymede. Jupiter is easily visible in the night sky (since it is brighter than Sirius) and finding it for the first time is very rewarding. Sometimes not all four moons may be visible, it depends on when you are observing.

Depending on how good your telescope is, you may be able to see more detail on the surface of the planet. Sometimes, you may even be able to see the shadow of a moon on Jupiter's surface. Do not get your hopes up. You will probably need at least a 6 inch telescope to really see any good detail or cloud bands. The great red spot is fairly difficult to see with an aperture less than 10 inches.

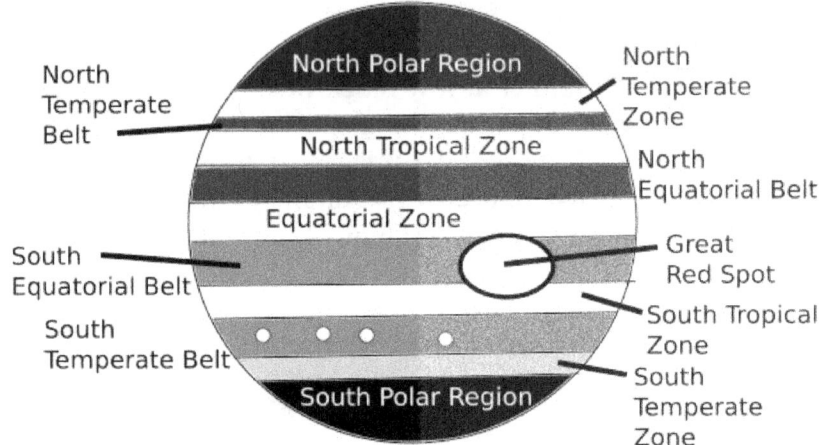

Figure 9.28 – Details on the surface of Jupiter.

I am able to see faint cloud bands using an 80mm refractor. I am sure it is possible with a 70mm refractor also. I captured the following image using my 90mm refractor.

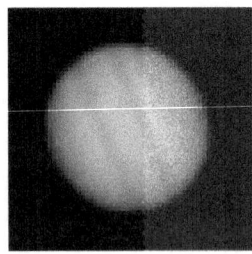

Saturn

Saturn is generally more difficult to observe than Jupiter. Saturn looks different from year to year. Saturn is as bright as a first magnitude star, but it is not as obvious as Jupiter or Venus in the night sky. Obviously, you want to see Saturn's rings. That is pretty much the goal of every amateur astronomer. Depending on the quality of your telescope, you may be able to see more detail, like the Cassini and Encke divisions. For small telescopes, the Cassini division is somewhat realistic, but the Encke division is nearly impossible to resolve.

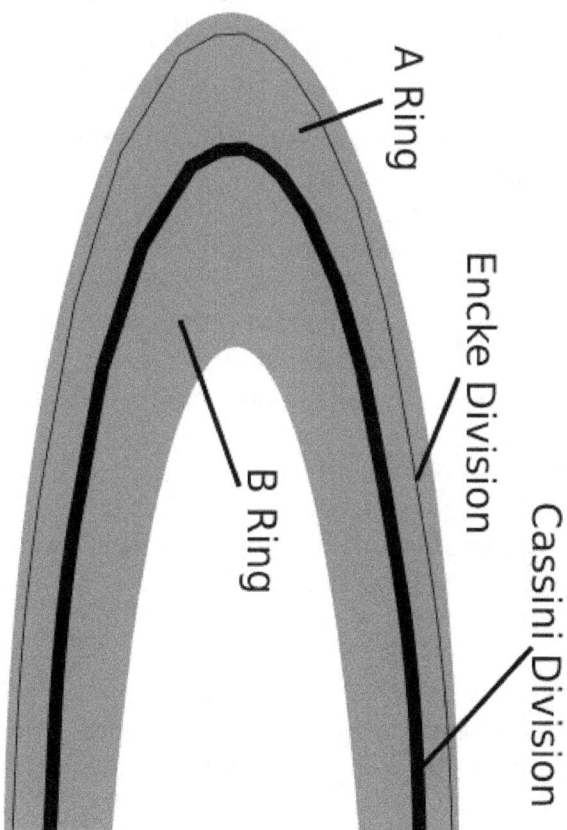

Figure 9.29 – The rings of Saturn.

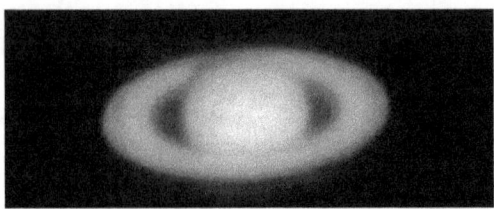

Figure 9.30 – Saturn at 130x through a 90mm refractor. Best 20 photos stacked. Poor seeing conditions.

Observing planets is not all too difficult of a task for the amateur astronomer. You just have to get a resource that tells you where planets will be at a given time. It is very fun to get together with some friends (or yourself), find a bright star, point your telescope toward it, and see if it is a planet or not. If you can find Sirius, and a "star" looks brighter than it, it is probably a planet.

To observe planets in detail, magnifications of ~200 times are needed. A 70mm refractor is probably the minimum aperture needed for detailed observations of Saturn and Jupiter (since a 70mm refractor can handle magnifications of up to ~200 times).

The only difficulty you will have observing planets is having the proper equipment. You need a quality telescope that can handle higher magnifications if you plan on seeing any detail.

Equatorial mounts are made to track the stars and DSOs, but you can use one to track planets for a short period of time. Planets move independently from the stars, but their motions are so similar in a small telescope over a short period of time that you should be able to track planets successfully using an equatorial mount.

The following three tables are included to show you where to look for Jupiter, Saturn, and Mars at certain times during the years 2020-2024. These are rough guidelines, so you should consult an outside source if you really want to know where to look. Many of these outside sources can be free, like a website or a smartphone application.

Where to find Jupiter

Year	Month	Where
2020	1-3	Morning
2020	4-6	Morning
2020	7-9	Evening
2020	10-12	Evening
2021	1-3	Not Visible
2021	4-6	Morning
2021	7-9	Afternoon
2021	10-12	Evening
2022	1-3	Not Visible
2022	4-6	Morning
2022	7-9	Morning
2022	10-12	Evening
2023	1-3	Evening
2023	4-6	Not Visible
2023	7-9	Morning
2023	10-12	Morning
2024	1-3	Evening
2024	4-6	Not Visible
2024	7-9	Morning
2024	10-12	Morning

Where to find Saturn

Year	Month	Where
2020	1-3	Morning
2020	4-6	Morning
2020	7-9	Evening
2020	10-12	Evening
2021	1-3	Not Visible
2021	4-6	Morning
2021	7-9	Afternoon
2021	10-12	Evening
2022	1-3	Not Visible
2022	4-6	Morning
2022	7-9	Morning
2022	10-12	Evening
2023	1-3	Not Visible
2023	4-6	Morning
2023	7-9	Morning
2023	10-12	Evening
2024	1-3	Not Visible
2024	4-6	Morning
2024	7-9	Morning
2024	10-12	Evening

Where to find Mars

Year	Month	Where
2020	1-3	Morning
2020	4-6	Morning
2020	7-9	Morning
2020	10-12	Evening
2021	1-3	Evening
2021	4-6	Evening
2021	7-9	Dusk
2021	10-12	Not Visible
2022	1-3	Morning
2022	4-6	Morning
2022	7-9	Morning
2022	10-12	Morning
2023	1-3	Evening
2023	4-6	Evening
2023	7-9	Dusk
2023	10-12	Not Visible
2024	1-3	Dawn
2024	4-6	Morning
2024	7-9	Morning
2024	10-12	Morning

Nature and the Sun

You should not underestimate how much fun observing nature with a telescope can be. A telescope is not limited to just the night sky. You can look at birds, squirrels, trees, or anything else outdoors. We will discuss the observing the sun on the next few pages. Amateur astronomers are very dependent on Mother Nature's cooperation.

Telescopes cannot look through clouds. That is a common misconception (and quite a strange one). If it is a cloudy night, we cannot use our telescopes (sadly). If the full moon is shining brightly, we cannot see some of the fainter DSOs (sadly). If the atmosphere is not steady, we cannot use high magnifications (sadly). If it is humid out and water gets on the optics of our telescopes, we might ruin our telescopes (sadly). During the summer, the constellation Orion and the Orion nebula will not be visible in the Northern Hemisphere (sadly). Some planets might not be visible at a given time (sadly) (although I am not sure Mother Nature controls the motions of the planets).

Everything we do as amateur astronomers requires full cooperation on Mother Nature's part. Most of the time, she just does not care about your silly telescope hobby. If she sees a low pressure area, she is going to fill it with a storm. If she does this, there is nothing you can do but wait for a clear night. Clear, steady nights might be quite rare depending on where you live.

It is quite frustrating and humorous just how "rude" Mother Nature can be. It seems that most of the time the Moon is near the new moon phase, it is cloudy outside; that most of the time there is a faint meteor shower, the full moon brightens the sky too much; and that most of the time you discover a new deep sky object you want to observe, it is not visible for another 3 months.

In order to observe the sun with a telescope, you need a proper solar filter. Please, do not use eyepiece solar filters. The solar filter you need must attach to the front of your telescope (between the objective lens and the sun). You can buy one, sure, but you can make your own very easily and save some money.

You do not need a large telescope to observe the sun. We do not need to capture any more light (the sun is very bright!) and the better resolutions associated with larger telescopes cannot be achieved during the daytime (the skies are not steady during the day). If you have a small telescope laying around, convert it into your solar telescope! Using a small telescope to observe the sun is a great idea for two reasons: (1) it gives you a reason to use the little guy again and (2) the smaller the solar sheet you have to buy, the cheaper the project is.

I decided to use my 50mm refractor as my solar observation telescope. I only needed to buy a 4"x4" solar filter sheet (solar filter sheets are cheap) and this was the only cost for the entire project. You also need a cardboard box and something like cardstock (thick paper or thin cardboard works too), but most of us probably have these laying around.

Learn how to build your own solar filter on the following pages.

Step 1: *Obtain the supplies. I used a piece of cardboard, some thin cardboard, a pair of scissors, a ruler, and a hot glue gun.*

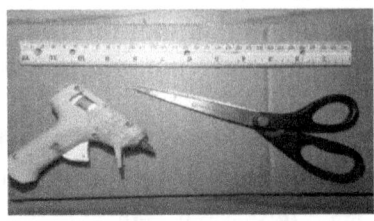

Step 2: *Design your filter holder. You need to draw two circles with an outer diameter larger than the outer diameter of the dew cap. You need to draw two smaller circles centered in the larger circles with a diameter close to the aperture of the telescope. It says 1.8 inches in the image, but I actually ended up using a 2 inch diameter. Use the cardboard for this.*

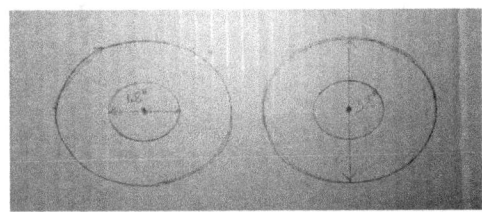

Step 3: *Cut out the circles. Take the cardstock and decide how wide you want the strip to be. I made mine 3 inches thick. Anything wider than 1 inch is probably sufficient.*

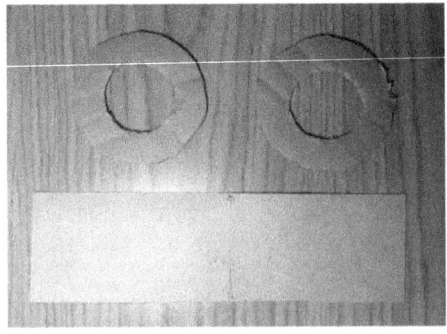

Step 4: *Wrap the strip around the dew cap of the telescope and mark where the overlapping edge ends. I backed away from this line a little so that it could slide on and off the dew cap easily. You want a tight fit so do not back off too much. I stapled my strip in place. You should be left with a cylinder with no top or bottom.*

Step 5: *Remove the finderscope from your telescope. You should be able to find the sun without a finderscope. It is not safe to leave the finderscope attached to the telescope.*

Step 6: *Attach the cardstock cylinder to one of the cardboard rings you made earlier. I used a hot glue gun.*

Step 6: *This step is "optional" but you really should do it. Look through the mount you have now in a bright room with your face pressed against the cardstock cylinder part. Do you see light coming through between the cardstock cylinder and the cardboard circle? If you do, you really should apply electrical tape or paint all round the area you glued so that no light is able to come through there. You only want light to come through the filter.*

Step 7: *The rest is relatively simple. All we have to do now is place the filter between the two cardboard rings and secure everything. You should try your best to align the inner circles perfectly with each other. The outer edges of the bigger circles do not have to be perfectly aligned with each other, but it is important the inner circles are aligned. Again, I used hot glue to secure everything.*

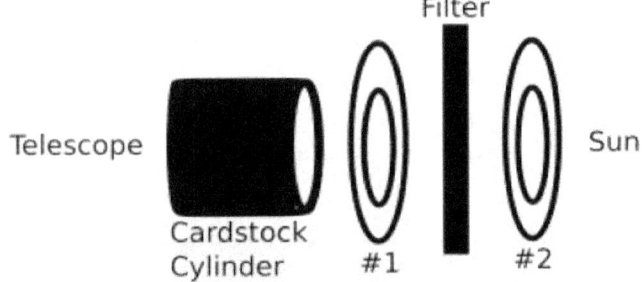

So how do we use this solar filter? Different types of filters show different details, but almost all filters show sun spots. Since you most likely bought a cheap filter sheet, sun spots are the best thing to observe.

Sun spots are "dark" regions on the sun that are still very, very bright. They are just dark relative to the other areas on the sun's surface. By the way, since the sun is a star, there really is no "surface". The photosphere is the first layer of the sun we can observe with a telescope (using a solar filter, of course). The chromosphere is outside of the photosphere, and the corona, outside of the chromosphere. The corona can be seen during a total solar eclipse or using a more expensive type of filter.

The sun goes through a sort of 11 year solar cycle, during which surface activity increases and decreases. During solar maximum, you can see many sun spots; during solar minimum, the sun may appear completely featureless. That's just how it works. Regardless of whether you are close to solar maximum or solar minimum, you should still observe the sun with your telescope. It is a great supplement to the normal nighttime observations.

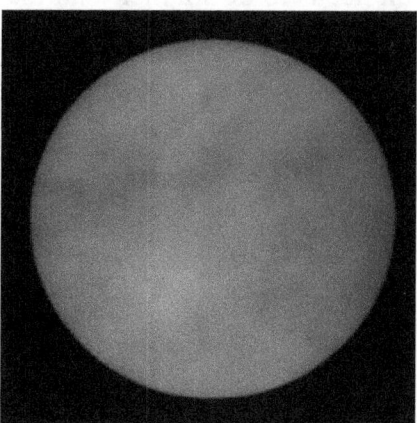

A small sunspot is visible, despite being close to solar minimum. This image was taken on a cloudy day (hence the clouds).

A great way to observe the sun with others is solar optical projection. You do not even need a solar filter, just some creativity and a 50mm refractor. You should not use an aperture greater than 50mm. Optical projection can be dangerous, so research how to do it safely and do it at your own risk.

(1) You need to cut a hole is a piece of cardboard and slide it over the dew cap of the telescope.

(2) Find something to use as a projection screen. A cheap white poster board from the local dollar store is fine. You need to figure out a way to secure the board to the ground.

(3) Remove the diagonal and attach an eyepiece directly to the focuser tube. You need to be "creative" here. If your focuser tube does not accept 1.25" accessories, go ahead and glue one on! That is what I did. Make sure to use a cheap eyepiece with low magnification. I used a 25mm eyepiece for my project. The sun needs to be able to fit into the field of view.

The further the telescope is from the poster board, the larger the sun will be. The sun needs to be smaller than the shadow from the cardboard light shield. The larger the shield, the larger the sun can be. You can also try this with the moon!

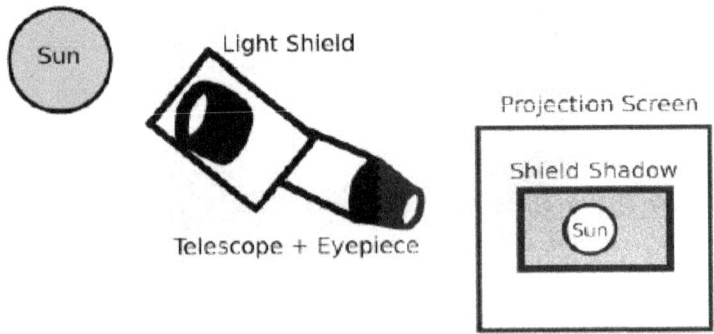

The following shots were taken during a partial solar eclipse that occurred on August 21st, 2017. They were taken using a 90mm refractor and the homemade solar filter created in this chapter.

Chapter 10

Astrophotography

It is one thing to see something through a telescope and talk to someone about it, but it is something else to get a picture of what you saw and show them (a picture is worth a thousand words). Professional astrophotography is *well* beyond the scope of this book, but you can still take decent pictures on almost any budget.

Why? Why even bother with astrophotography? Well, it certainly is <u>not</u> a way to make easy money. Plenty of people take on astrophotography and excel at it. In addition, no matter how much money you spend, you can never beat the Hubble space telescope. Astrophotography is simply a fun, challenging hobby. In my opinion, it is quite rewarding to create these images yourself.

I also don't understand the reluctance for very good amateur astronomers to sell their work. I bet billions of photos exist of the Eiffel tower. Some photos taken by professionals and some taken by horrible cell phones. Still, people photograph the Eiffel tower and I bet people still make money selling their photos of the Eiffel tower. Astrophotography is photography and people will always pay for photos taken by good photographers! Just because the Hubble has taken pictures of M31, doesn't mean you shouldn't and just because the Hubble photo is "free" doesn't mean people won't pay you for your version.

There are a few ways you can go about taking these pictures. I am going to break up the rest of this chapter into 7 parts:

(1) Smartphone astrophotography without a telescope
(2) Smartphone astrophotography with a telescope
(3) DSLR astrophotography without a telescope
(4) DSLR astrophotography with a telescope (Lunar/Planetary)
(5) DSLR astrophotography with a telescope (DSOs)
(6) Dedicated Astronomy Cameras (DSOs)
(7) Dedicated Astronomy Cameras (Lunar/Planetary)

Aperture is a feature of the lens (or telescope) you are using. The aperture is given like the f-number with telescopes. Generally a larger aperture (lower f/#) will yield a brighter image for a given shutter speed and ISO. Aperture effects the depth-of-field (DOF).

Depth	Caused by...	Result...
Small	Large apertures (f/2)	Blurry secondary objects
Large	Small apertures (f/10)	Most of image in focus

Confused by this table? If not, skip this paragraph. Consider a table with 3 objects on it. Object 1 is closest to you and object 3 is furthest from you (object 2 is between them). If you set a large aperture (small f/#) and focus on object 2, objects 1 and 3 are likely going to be out of focus. This is sometimes a desirable effect. If you set a small aperture (larger f/#) and focus on object 2, objects 1 and 3 may also be in focus.

What does this mean in astrophotography? If you are interested in taking star bokeh photography (more about this later), then you need to use a large aperture.

In astrophotography, we generally want a large aperture so that we can capture as much light as possible. If only it was that easy... Some lenses, actually most affordable lenses, have some troubles at large apertures. For example, stars near the edges of the photographs may have wings or appear severely distorted. To fix this, we make the aperture slightly smaller. In general, f/2.2-f/3 is usable but you will have to do your own tests with whatever lenses you have.

Of course, if you attach a DSLR to a telescope, you have no control over the aperture!

Shutter speed refers to the time the camera sensor is accepting photons. In astrophotography, we enthusiastically scream "BRING ON THE PHOTONS!!!" We want a lot of photons in our images. But, of course, it is never that easy.

The earth rotates right? This means the night sky rotates. And this means that the stars move. And this means that the DSOs, planets, and moon move. If our shutter speed is too long, our photo will have obvious streaks. See the following example.

30 seconds without Barn Door Mount

30 Seconds with Barn Door Mount

Ignore the "barn door" thing for now. Look at the first image. We see streaks and, in some cases, stars completely faded away due to the streaking. Not good. If you have a form of "tracking", you can take longer exposures (longer shutter speeds). We will discuss different ways to "track" but clearly the second image shows that "tracking" helped a lot!

Shutter speed limits change a lot depending on the type of photography you are doing. Therefore, we will discuss this in more detail later.

ISO. This is the big one. This is also the one that causes the most arguments. Essentially, higher ISOs simply "boost" or "amplify" the brightness of the image. A very, very common misconception is that higher ISOs introduce more noise into the image. (I must admit, sometimes I find this to be true but do not tell anyone). In my opinion, the best way for YOU to tackle this ISO problem is to experiment. If you don't feel like experimenting, stick with ISOs around 1600 as your maximum. Newer cameras will likely be able to handle higher ISOs better than older cameras.

General Photography

As a bonus, let's discuss DSLR or mirrorless photography quickly. In general, the workflow is simple:
1. Set your camera to Av (aperture priority) mode or M (Manual) mode.
2. Set your aperture to get the desired DOF (depth of field) you want. You want a blurry background? Set as low as you can go.
3. Set your desired ISO, and adjust your ISO as needed to get a neutral exposure.
4. In Av mode, watch the shutter speed the camera returns.
   a. If the shutter speed is too slow, increase your ISO. If you are at your ISO maximum, use a larger aperture.
   b. If the shutter speed is unnecessarily fast, decrease your ISO

Photography, put very simply and generally, is all about balancing aperture, ISO, and shutter speed.

Smartphone Astrophotography (no telescope)

This is the cheapest method we will discuss. All you need is a smartphone and an application capable of taking long-exposure photographs. You can get free ones, but I suggest dropping 3 dollars on a quality one.

You can choose to use a tripod or not to. Any tripod will work, but you will need to get an adapter. This adapter will make it possible to mount a smartphone on any normal tripod. These adapters are usually inexpensive. You do not need a tripod. Get a sandwich bag and fill it with dirt, sand, or even dry rice. You can rest you smartphone on that bag and take the photos. That bag will keep your smartphone steady.

Understand something here. You are using a smartphone camera. Smartphones have good cameras, but they are not that great for astrophotography. You do not want to zoom in. At all. You will already have a lot of noise in the image without zooming in.

This is actually a good thing. Since you will be so "zoomed out" the objects (stars, DSOs, etc...) you are shooting will be essentially stationary for a longer period of time. This means you can actually take pretty long exposures with your smartphone. I like to take long-exposures of 15 seconds with my smartphone. I see no streaks when I do. You could take longer exposures, but you need tracking to do so.

For ~5 dollars you can take pictures of the night sky (if you already have a smartphone and use the sandwich bag trick). Obviously, the results will be much better in dark skies. I do not recommend this method. If you want to take pictures of stars, you need a "real" camera. A smartphone camera, however, can be used with a telescope for decent results.

Long-exposure (15 seconds). Smartphone.

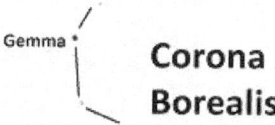

Corona
Borealis

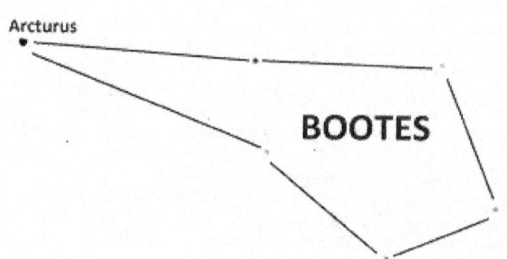

BOOTES

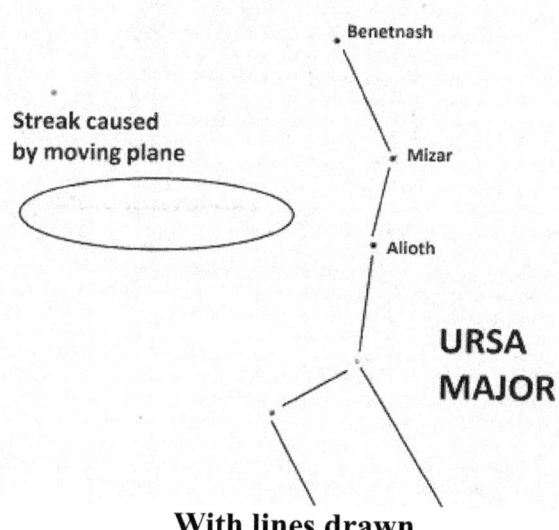

With lines drawn.

Long-exposure (15 seconds). Smartphone.

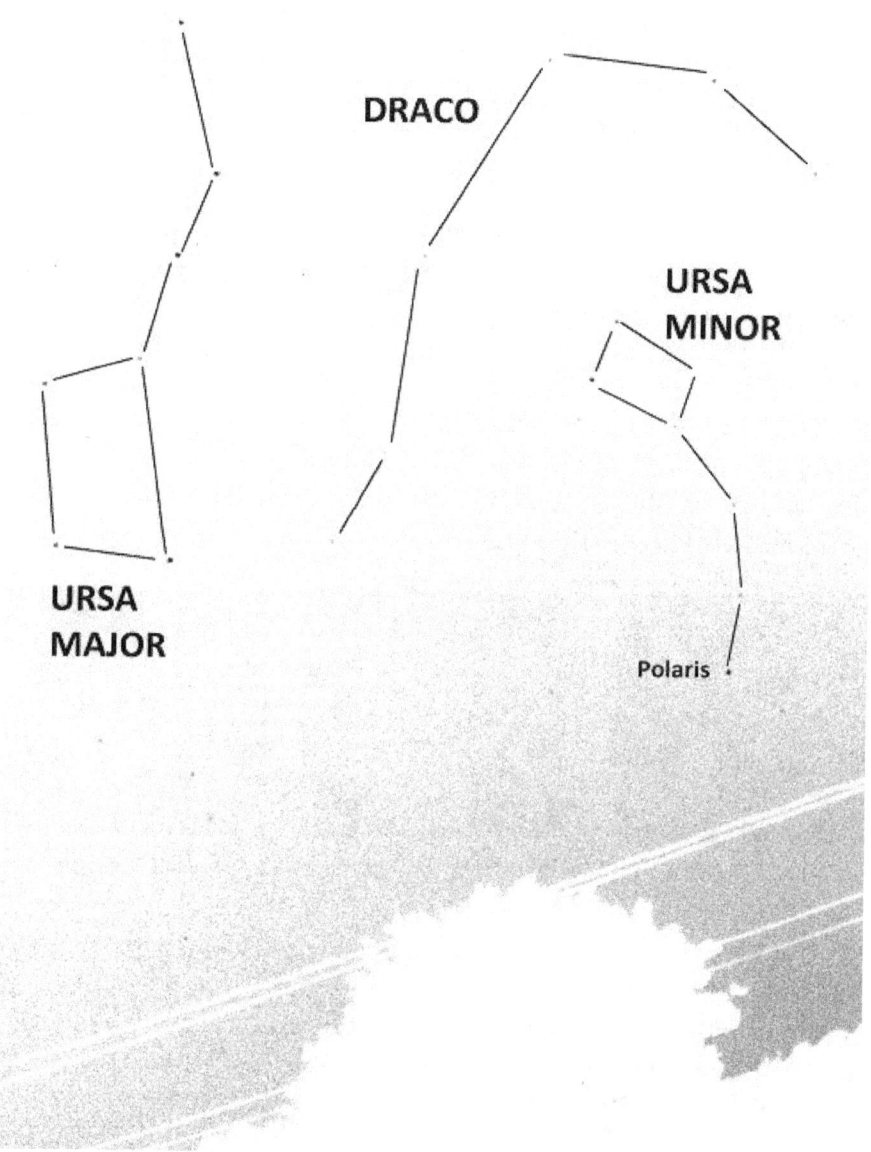

DRACO

URSA MINOR

URSA MAJOR

Polaris

With lines drawn.

Smartphone Astrophotography (with telescope)

You can also use your smartphone with a telescope. When you do so, you can take pictures with your smartphone at higher magnifications. The easiest way of doing this is getting a universal mount that connects any eyepiece to any smartphone. These are inexpensive, unless you get an extraordinarily good one. I use a low-quality version and it works just fine.

These mounts can be frustrating though (although not as frustrating as the afocal method). The smartphone camera must be perfectly aligned with the eyepiece. To make sure everything is aligned properly, you must open your smartphone's camera after installing the eyepiece, the mount, and the smartphone to the telescope (make sure to take off the caps). Now go inside in a bright room and aim at a bright wall or shine that red flashlight onto a piece of paper laying across the dew cap and tweak the installation until everything is aligned properly.

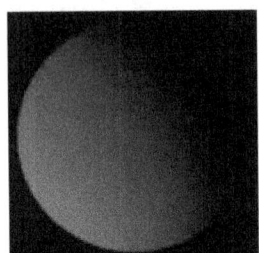  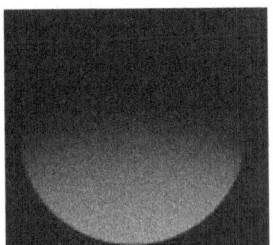

Not Aligned   _Aligned Properly_   _Not Aligned_

When everything is properly aligned, the edges of the circular field of view will be crisp and clear. Since you care going to want to change eyepieces over the course of the night, you are probably going to want to learn how to do this outside in the dark (using the red flashlight method).

Once the camera is all lined up and ready to go, you can go outside and start taking photos. I like to have headphones with volume controls plugged into the phone. That way, I can simply click the volume button to take the picture and not shake the telescope by touching the phone.

Alternatively, you can set up your camera to take the picture a couple seconds after you touch the phone. This way, the telescope will settle and stop shaking before the phone takes the picture. Smartphone astrophotography has its limitations, but it is relatively inexpensive (if you have a smartphone) and simple.

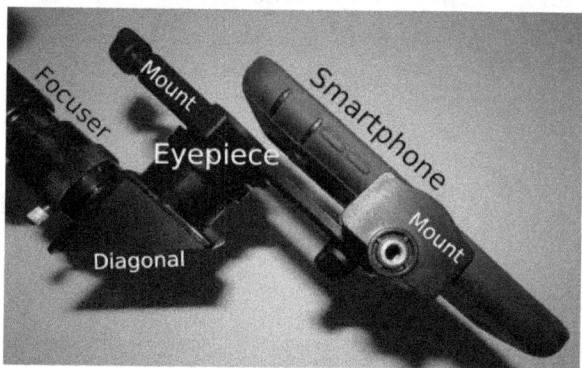

Figure 10.1 – Using a smartphone mount to take photos.

Since you are taking the pictures at higher magnifications, you cannot take long-exposures without accurate tracking. You need an altazimuth stand with two motors (declination and right-ascension) or an equatorial stand with a RA motor. Accurate tracking is very expensive. Buy a DSLR camera and get used to it before you spend a fortune on accurate tracking. There is no point in having accurate tracking without a "real" camera.

It is usually relatively easy to take pictures of the moon, planets, and stars with a smartphone, but fainter objects can be more difficult to capture. If you are trying to get an image of a fainter DSO, it is best to do so when the moon is not up and during dark and steady skies.

I have included a good example of this problem below. I took this photograph of M13 using my 90mm refractor and 23 times magnification. At the time, the moon was up and I was 20 minutes away from the city of Philadelphia (in other words, the sky was not dark). To make matters even worse, M13 was in the direction of the city. If there is a major city to the west of your location, objects in the western sky will appear brighter if you wait for them to move into the eastern sky (since the background sky will be darker in the east).

Figure 10.2 – Photo of M13 (inverted to save ink).

If you are not impressed with this image of M13 (you should not be) take a look with your own telescope. It will look brighter and better than this.

M41. Smartphone with 90mm refractor.

In an attempt to avoid the cold, you may try to use your telescope indoors and observe through a window. While this may seem smart, the refraction of the light caused by the difference in temperature and the window glass will alter the image. That being said, it is still possible (but not recommended). I took the image in figure 10.3 through a window using my 50mm refractor. If you are going to do this, at least make sure all the lights are out in the room.

Figure 10.3 – Photo taken through a window.

DSLR Astrophotography (no telescope)

This is where astrophotography becomes fun. Unfortunately, this is also where astrophotography can become expensive and difficult. You will need a few things to pursue astrophotography using a DSLR without a telescope:

(1) **DSLR Camera.** You need a DSLR camera. These can be *very* expensive. If you have no budget, go ahead and spend a fortune on a high-quality DSLR camera. If you are on a budget (like me), then, believe it or not, there is a way! Buy a heavily used DSLR camera. When I say heavily used, I mean *heavily* used. Find something with a lot of exterior wear, a broken flash, and no accessories. You should be able to get a used DSLR camera for a good price at a wide variety of used camera stores. Please just make sure the sensor is free of scratches.

Do not fear a beaten up DSLR. The images you get will still blow you away. When looking for a DSLR, make sure it can do long-exposures and that you can change some settings, like ISO. I would imagine that every DSLR camera has these features, but make sure. I have tried to avoid typing the words of a specific brand throughout this book, but any Canon DSLR will work perfectly. Get a Canon if you do not have a DSLR already.

I bought my DSLR heavily used. It came with significant exterior damage, a broken flash, no lens, a clear sensor, a battery, a charger, and nothing else. Perfect! All I needed was a lens, SD card, and a reader to download the images onto a computer.

I want to talk about megapixels. Megapixels are over-rated. 8 megapixels is just as good as 10 pixels unless you are trying to severely blow up the image. Do not let megapixels force you to buy a more expensive camera. A 2 megapixel difference is not going to significantly change anything. My DSLR has 8 megapixels. It is fine for beginners.

(2) **A lens**. Camera lenses are also very expensive. Solution? Buy a heavily used camera lens. This may seem funny, but if you are on a budget, this is absolutely the way to go! The lower the f/#, the better.

The focal length you want depends on what you want to shoot. If you want to take wide-field shots of the Milky Way and constellations, you should aim for a focal length close to 25mm. The lower the focal length, the lower the magnification, and the longer the shutter speeds you can use before stars start to streak. If you want to shoot pieces of constellations and some of the larger DSOs, use a longer focal length lens (~100mm). You could also just use a telescope and a DSLR, which we will discuss shortly. In that case, the focal length of the telescope's objective lens is equal to the focal length of the camera "lens".

(3) **Tripod.** Unfortunately, you are going to want a tripod for a DSLR. Since you are using an expensive camera (regardless of how much you paid for it), it is recommended you get yourself at least a decent tripod. The tripod you received with your telescope might also work with your DSLR camera. Check to make sure before you buy a tripod. You may also be able to mount your camera on top of your telescope. You want a sturdy tripod!

(4) **Remote Shutter.** You do not want to touch the camera as it takes a picture. The shaking will ruin the image. Remote shutters make it possible to take a picture remotely. You can get wired ones or wireless ones. I like the wired ones because they do not require batteries. If you plan on taking any "bulb" exposures (more on this later), you will definitely need a remote shutter. You may want to consider purchasing an intervolameter. This will ultimately make star trail and deep sky object photography easier.

Recommended DSLR Camera Settings for wide-field:

White Balance: If your camera is set to RAW mode, white balance does not matter because you can change it during the editing phase. I find that "daylight" is the best for accurate star colors. If you must shoot in ".jpg" use whatever looks right. You can also use certain photographic gray cards to set a proper white balance.

Aperture: Also, f/# setting. Set this to the lowest value you can. However, as we mentioned, this may cause some problems. If you see some strange deformities near the edges of the photos, decrease the aperture (increase the # in f/#) until you are happy. This will result in a darker, noisier image so be careful.

ISO: The best ISO for your camera depends on your camera. You really just need to experiment. However, I can offer the following starting points. For the Milky Way, use the highest you can (3200 would be very helpful!). For star trails, I would try 400-800. For general constellation photos, try 800-1600. Obviously these values are heavily dependent on the apertures and shutter speeds you use.

You should only try lunar and planetary with a telescope.

Shutter Speed: You either have a full frame DSLR or an APS-C DSLR. If you have a full frame, you can probably get away with using the "500 divided by the focal length rule" for the longest exposure before stars start to streak. If you have a APS-C camera, you can probably get away with using the "300 divided by the focal length rule" for the longest exposure before the stars start to streak (for example, using 18mm lens gives 16.67 seconds of maximum exposure time). Objects near the pole will streak slower, so you may be able to use longer exposure times. Objects further from the pole, may start to streak even using these rules! See the following table.

Lens	Full Frame	APS-C
10mm	50.0	30.0
12mm	41.7	25.0
15mm	33.3	20.0
17mm	29.4	17.6
20mm	25.0	15.0
22mm	22.7	13.6
24mm	20.8	12.5
26mm	19.2	11.5
28mm	17.9	10.7
30mm	16.7	10.0
35mm	14.3	8.6
40mm	12.5	7.5
45mm	11.1	6.7
50mm	10.0	6.0
55mm	9.1	5.5
65mm	7.7	4.6
75mm	6.7	4.0
85mm	5.9	3.5
100mm	5.0	3.0
120mm	4.2	2.5
150mm	3.3	2.0
175mm	2.9	1.7
200mm	2.5	1.5
300mm	1.7	1.0

Table 10.1 – Maximum exposure lengths (seconds).

What about filters and attachments?

Your lens has a number printed on it and that number determines the types of filter or attachments you can screw on. If you see 58mm, you can buy 58mm filters and attachments. If you have a 52mm lens and want a 58mm filter, you can purchase a 52mm to 58mm step-up ring. Step down rings are a bad idea and may cause dark corners.

You will find items like 0.4x and 2.5x attachments. These are supposed to decrease or increase, respectively, the magnification of a lens. In general, you want to minimize the amount of optics between your camera and your desired target. It is a well-accepted belief in astrophotography that cropping the image 2x is better than a 2x attachment. A lens hood is a good way to eliminate stray light!

Filters are a good idea. Forget UV filters. In my opinion, there are only 2 filters you should consider and one of them is just for artistic purposes.

(1) Red intensifier filter. This filter is a very affordable way to limit the effects of light pollution.

(2) Fog filter. You may have come across photographs that have enlarged stars that emphasize their color. While some do this in post-processing, you can get this effect using a fog filter.

The following image was produced using a 167 second exposure and both a fog filter and a red intensifier filter. Vega looks huge!

Star trail photography is very simple and looks very impressive. It takes advantage of the fact that the Earth rotates and the stars move throughout the night. See the example below.

Pretty cool, right? This is scenario where an intervolameter comes in handy. To make things easy, I will let you know the settings I almost always use. I use ISO 400, shutter speed of 30 seconds, aperture f/2.8, and 50mm focal length. Of course, this is for my equipment. Simply take a bunch of these photos.

I set up the intervolameter to take 30 second exposures with a 1 second interval for as many frames as I want (I prefer 60-100 frames for processing time reasons). Upload all of the frames into a photo editor as layers and change the mode to "lighten" for all of them. That's all there is to it. Flatten the image before editing it to save a lot of time.

If you have a lens with a larger aperture (~1.8-3.0) you can take star bokeh photographs. They are fun to take and tend to impress people. Simply focus on the foreground and let the stars remain out-of-focus. I have included two examples.

DSLR Astrophotography – The Big "Problems"

There are 3 major problems with DSLR astrophotography.

(1) The Tracking Problem: EQ mount vs. Barn Door Tracker

The stars move. We have already covered that. Landscape (trees, mountains, etc.) does not move. If you have landscape in your photograph, tracking will keep the stars looking good but will blur the landscape! You do have an option though. You could take one image for landscape "foreground" and take another image for the stars and blend them together using a photo editing software.

Regardless, in order to take really great photos of the night sky, you need tracking! Longer exposures create images with much less noise because more data (signal) has been collected. Some of you may be familiar with the term "signal-to-noise-ratio" or SNR. Longer exposures collect more signal, resulting in a higher SNR! ISO is not related to this.

So, how do we track? We have two options. We will discuss the barn door tracker mount first. However, you are probably just going to want to buy an equatorial mount!

Barn Door Tracker

Perhaps the pinnacle of budget astrophotography is the barn door tracker mount. It is a fun DIY project and can actually give you decent results. Only use this mount with short focal lengths (less than 75mm). There is plenty of information out there on these but I will show you how to build one. A diagram can be found on the next page.

The concept is quite simple. If you align a door hinge with the northern pole and swing the camera around this "axis", the camera will track the stars.

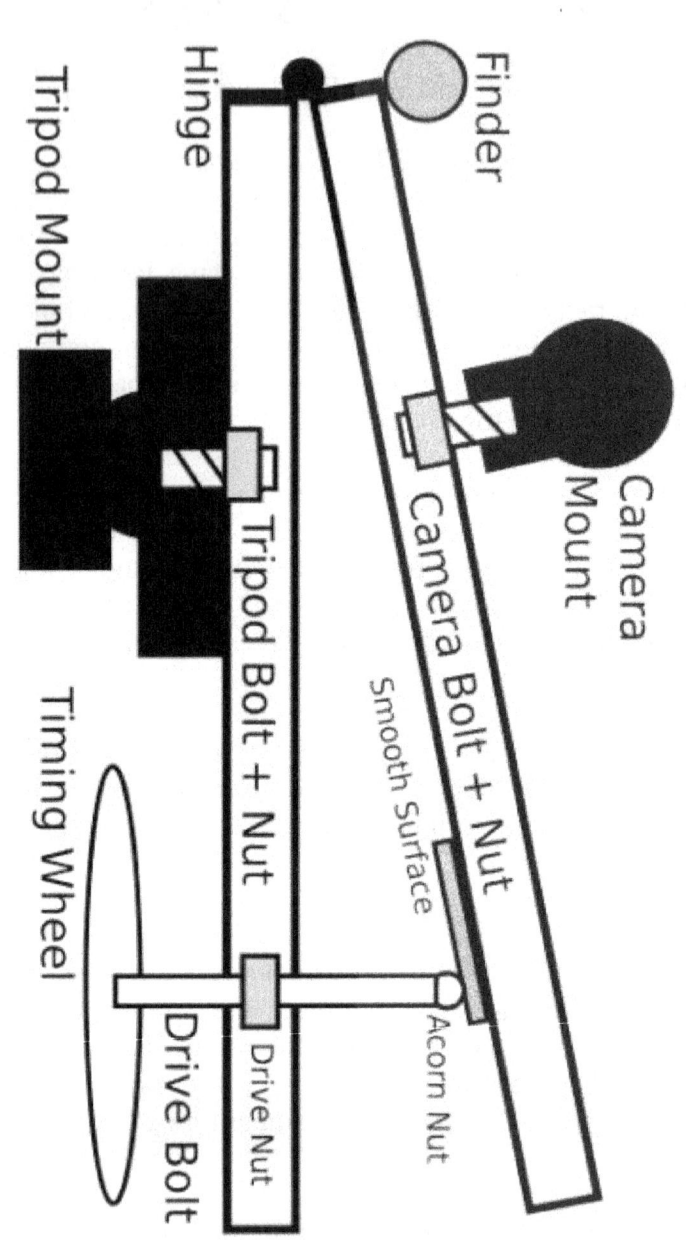

Finder

Hinge

Tripod Mount

Camera Mount

Tripod Bolt + Nut

Camera Bolt + Nut

Smooth Surface

Timing Wheel

Drive Bolt

Drive Nut

Acorn Nut

You have some design choices.

(1) Do you want to spin the drive bolt manually? You should if you are doing this on a budget. If so, attach some type of timing wheel. The rate at which the timing wheel spins is the same as the rate the drive bolt spins.

(2) What drive bolt do you want to use? You have to consider the size of the bolt and the threads-per-inch (TPI). You are free to choose your own but I suggest using a TPI of 24 or 32. The drive bolt I chose for my following design was a 12 inch long 10-24 bolt.

(3) What RPM do you want to use? If you are going manual I suggest 1.5 RPM. Why? The timing wheel is easy to design.

Once the drive bolt and RPM are chosen, you have to do some math. We have to calculate how far the drive bolt has to be from the hinge. The following distance is the distance from the center of the hinge from the center of the drive bolt. Be accurate.

$$Distance\ (inches) = \frac{RPM}{(0.004375 * TPI)}$$

(4) What is going to be your tripod and what is going to be your camera mount? You have to be able to move the tripod head so that you can align the hinges with the northern pole. Once the hinges are aligned, you cannot touch the tripod again. The hinges have to be aligned with the northern pole the entire time you are imaging. This is why you need some type of camera mount (you need to be able to point your camera at your desired target!). Some type of ball joint camera mount is perfect. Tripod and camera bolts and nuts are usually ¼" and 20 TPI (1/4-20). I used an altazimuth mount as my tripod. I used the top portion of a camera tripod as my camera mount. Yes, I sacrificed a cheap tripod for this project.

Back to the manual wheel. You have to spin this wheel at a precise rate to keep the camera aligned with the stars. This is actually easier than it sounds. Say you are using a 50mm lens. With an APS-C camera, the stars will appear stationary for approximately 5 seconds. In this case, you only have to adjust the timing wheel every 5 seconds. Again, designing the wheel is easy if you choose to use a 1.5 RPM spin rate.

First, mark the center of the timing wheel and then draw two perpendicular lines (as shown in the image).

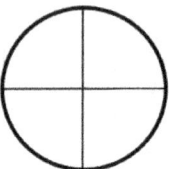

Second, draw four 45 degree lines, splitting the four sections into 8 sections.

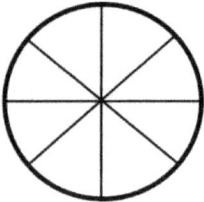

Finally, draw 9 degree tick marks (use a protractor). One section is done in the following image. Since 45 is divisible by 9, four tick marks are found in each section.

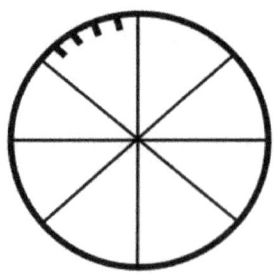

I skipped all the boring math here, but what exactly does this mean? Since we chose 1.5 RPM in our design plans, we have to turn the wheel 1/8th of a turn (45 degrees) every 5 seconds. You could also think of it as turning 9 degrees every 1 second.

If you are using a 50mm lens, you have to turn the wheel 45 degrees (1 "section") every 5 seconds (since 5 seconds is the maximum shutter speed before stars start to streak). In reality, we always have to turn the wheel 9 degrees every 1 second but a 50mm lens never sees the benefit since it is so zoomed out.

Barn door mounts work well, but you may have a difficult time getting even 30 second exposures without streaks. Your build has to be sturdy. If you shake the drive bolt, you may shake the camera!

Like most things in astrophotography, barn door mounts are as expensive as you want them to be. But once you start spending around a hundred dollars, you might as well save up for an equatorial mount with RA motor drive.

Learn how to build your own single-arm barn door tracker mount on the following pages. I will explain how I built mine.

Materials:

(2) 5.5 in. x 0.5 in. boards of wood (2 feet long each).
(2) Hinges with screws.
(1) Strong adhesive. We will be bonding wood and metal.
(1) Drive bolt. 12 in. long 10-24 bolt.
(1) 10 – 24 acorn nut.
(1) 10 – 24 tee nut (drive nut).
(1) Timing wheel with nuts and washers that fit.
(1) ¼ – 20 tee nut (tripod nut).
(1) Large washer (flat surface) with bolt and nuts that fit.
(1) Altazimuth Tripod.
(1) Camera Tripod.

Instructions:

(1) Design the boards as shown. Drill 1/4" or 5/16" holes for the tee nuts. Hole for flat surface bolt should be size of bolt threads.

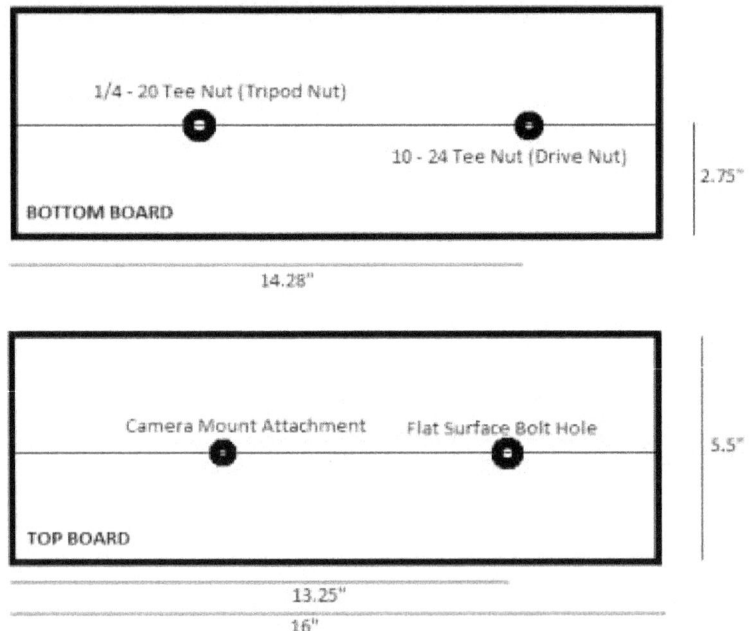

(2) Insert the parts. Hammer the tee nuts into the base of the wood. Glue the camera mount to the top board. Make sure the camera mount can still swivel. You may wish to glue the nut/bolts and other parts and pieces as you go.

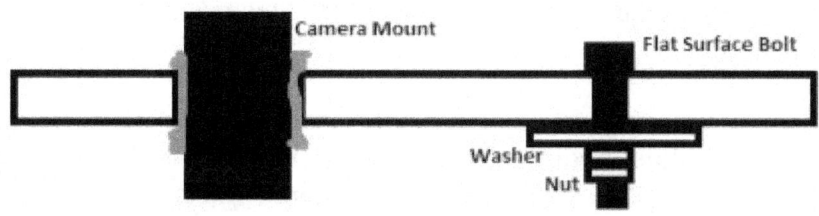

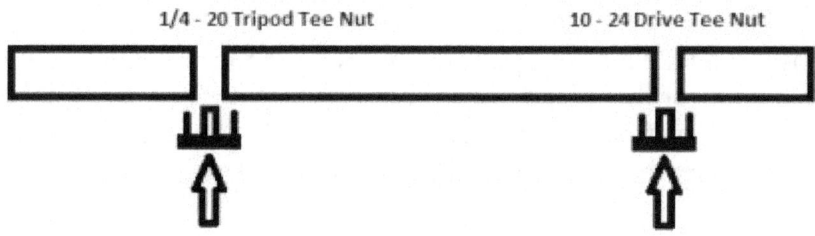

(3) Final product.

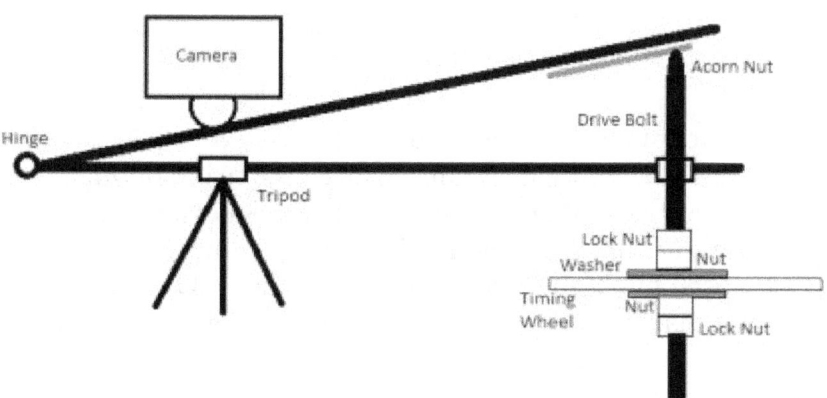

Equatorial Mount

Despite the fact that building your own barn door mount is a great DIY project, it is almost a waste of time. You will get much more accurate tracking with an equatorial mount. If you use focal lengths shorter than 100mm, even a very cheap equatorial mount will be sufficient.

An equatorial mount is a complex instrument. It even has its own chapter in this book! Please refer to that chapter for more information on equatorial mounts. In general, you are going to want to get an equatorial mount at some point. They are the only way to get accurate tracking!

(2) The Exposure Problem

When you take a photo, you will likely have the ability to view the histogram for the image (do not be scared of it). You can probably reveal this histogram by pressing the "display" or "info" button a few times on the photo. You will probably see one of the following.

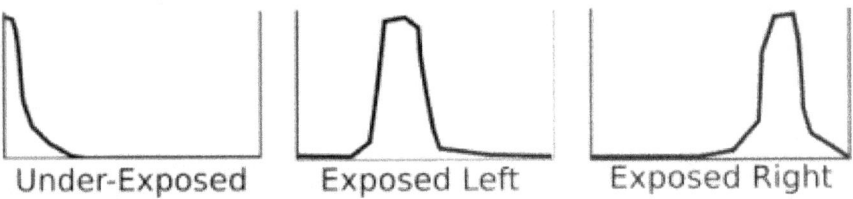

Under-Exposed   Exposed Left   Exposed Right

If your histogram resembles "under-exposed", your image is going to be noisy. You will have to increase the exposure during editing and introduce a lot of noise into the image. You want to avoid this histogram. How do we avoid it? We need to either (1) use a longer shutter speed or (2) use a lower f/# (assuming ISO is maxed out). Since (1) may introduce star trails into our image, we have to pursue (2). Since (2) probably costs a lot of money, you can see why many people invest in some type of tracking method.

If your histogram resembles "exposed left", you may be okay. Editing may introduce a little noise into the image, but you will probably be pleased with the image.

If your histogram resembles "exposed right", that is incredible. It may appear overly-bright in the camera, but you can easily decrease the exposure during editing. This is the best for a low-noise photograph. Just make sure you do not expose too long! If any piece of the curve is missing, you lost data. You want the curve to be away from both the left (under-exposed) and right side (over-exposed) walls.

(3) The Focus Problem

Focusing a DSLR camera for astrophotography is not as trivial as you would like. These camera cannot auto-focus the stars, so you will have to do so manually. On your lens, you will see a switch labeled AF (auto-focus) and MF (manual-focus). Set the switch to the latter.

Stars must be points. Sure, in theory you could look through the camera's viewfinder and focus the lens until the stars look like points, but this is a very frustrating task. Alternatively, you could do this roughly and take photographs, changing the focus slightly between exposures to narrow down the best focus point. This is all very frustrating though.

Solution? Get a camera with "Live View" or something similar. This feature will allow you to look at the screen to focus the stars. Additionally, this feature will likely allow you to zoom in on a portion of the screen. If you have this feature, you can easily get the best focus on a star. In my opinion, you should get a camera with some sort of Live View function. It is almost mandatory.

A very useful trick is called a star focus mask. Most popular is the Bahtinov mask. Look it up! This mostly applies to telescopes.

Things get very messy here. You can do afocal (do not do afocal), prime focus, or eyepiece projection.

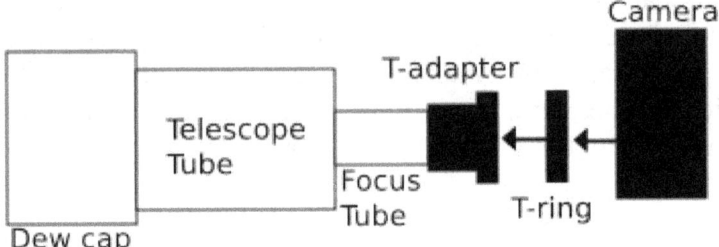

Figure 10.4 – The prime focus method.

The prime focus method requires you to have the following:

(1) DSLR Camera. See previous section.

(2) Remote Shutter. See previous section.

(3) T-ring and T-adapter. You do not need a camera lens for this method, because your telescope is the camera lens. The focal length of the telescope is the focal length of the camera's "lens". To connect a telescope and a DSLR camera, you need the appropriate T-ring and T-adapter. A T-ring is made to work with your DSLR where you screw in a lens. A T-adapter has the barrel that works with the focuser tube or diagonal and attaches to the T-ring. The T-ring has to be compatible with your camera and the T-adapter has to be compatible with your telescope and your T-ring.

(4) Refactor telescope. Of course, you could use a reflector too with very minor differences.

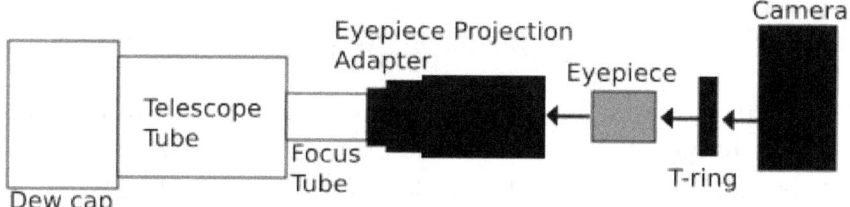

Figure 10.5 – The eyepiece projection method (using an eyepiece projection adapter).

The eyepiece projection method requires the following:

(1) DSLR Camera. See previous section.

(2) Remote Shutter. See previous section.

(3) T-ring. See previous section.

(4) Refactor telescope. Of course, you could use a reflector too with very minor differences.

(5) Eyepiece.

(6) Eyepiece projection adapter. You should not have a problem finding one of these for an affordable price.

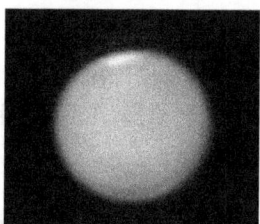

Photo of Mars taken using the eyepiece projection method.

The moon is a very bright target. If you just want to take a single shot of the moon, simply aim your telescope and fire away.

The prime focus method is mostly used to capture the entire moon in one frame. Use ISO around 100-400 and adjust shutter appropriately. I usually use around 1/200s. You should enable "mirror lockup" to avoid blurry images (usually found under custom functions). Use a remote shutter.

For refractors: you can add a Barlow lens between focus tube and T-adapter to slightly increase magnification. This may require the addition of a diagonal.

For reflectors: you may not have enough back-focus to achieve focus using the prime focus method. You can (1) add a Barlow lens to see if this helps you achieve focus or (2) cut a few inches off the bottom of your telescope tube (where the mirror is located). There are other fixes to this problem if you do not feel comfortable dissembling and cutting your telescope! If you do cut a few inches off your telescope, you may lose the ability to use your telescope visually!

The eyepiece projection method usually disappoints me but it is the best way to get close photos of individual craters and regions. The settings you use in your camera are greatly affected by the setup and magnification you are using. All you can do is experiment.

Regardless of the method you use, focusing your telescope is relatively easy. If you have some sort of Live View functionality, enable that and zoom in on a crater. Adjust the focus until it appears perfect!

In general, lunar photography is very simple. But, we can make it very complicated if we wish. If you want to get high quality images of the moon, you are probably going to end up taking videos and processing them.

Photograph of the moon taken using prime focus method.

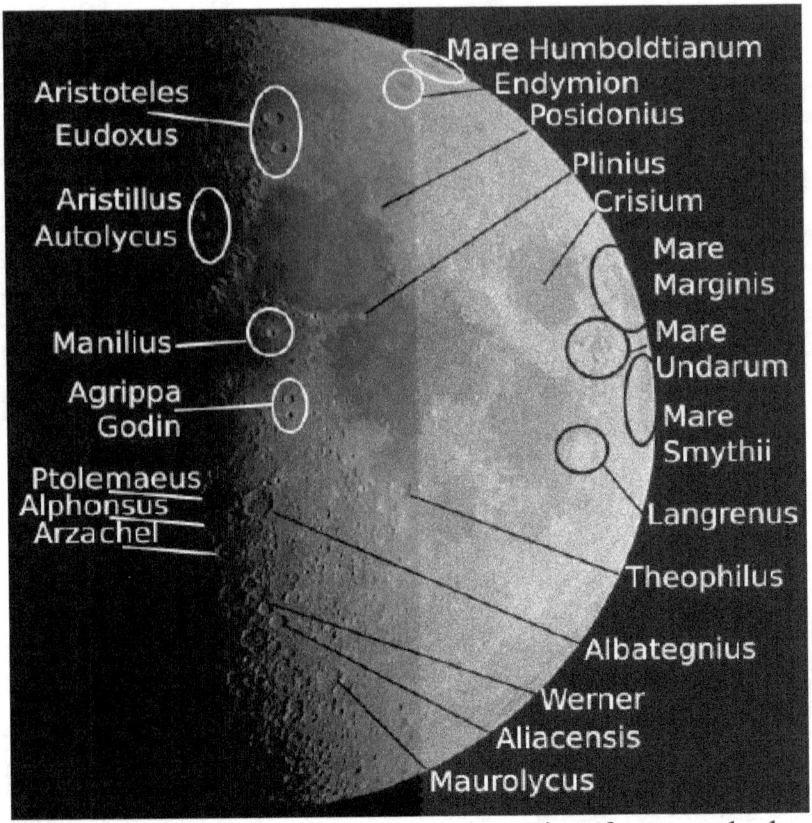

Aristoteles
Eudoxus

Aristillus
Autolycus

Manilius

Agrippa
Godin

Ptolemaeus
Alphonsus
Arzachel

Mare Humboldtianum
Endymion
Posidonius
Plinius
Crisium
Mare Marginis
Mare Undarum
Mare Smythii
Langrenus
Theophilus
Albategnius
Werner
Aliacensis
Maurolycus

Photograph of the moon taken using prime focus method.

Photograph of the moon obtained using a webcam video.

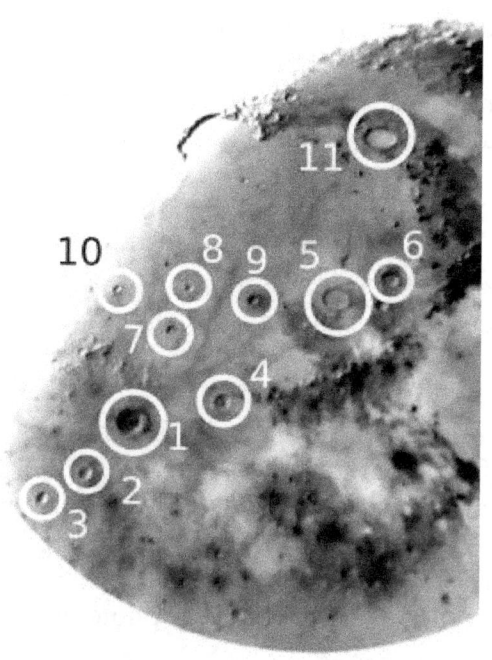

The moon. Smartphone with 90mm refractor.

1	Copernicus
2	Reinhold
3	Lansberg
4	Eratosthenes
5	Archimedes
6	Aristillus
7	Pytheas
8	Lambert
9	Timocharis
10	Euler
11	Plato

Currently, there are two common ways to take budget-friendly videos of the moon and process them. (1) You can use a webcam or (2) you can use a DSLR. DSLRs are not ideal for this application unless they have some type of "crop mode".

For images of the entire moon or large regions, most people are happy with prime focus, single-shot photography. Videos are usually only taken for small regions and individual craters. If you do take a video at high magnifications, you may need tracking or the video will only be a few seconds long.

Planetary photography is related to lunar photography in some ways. Frist of all, both targets are bright and do not require long exposures. You may need tracking, however, if you plan on taking longer videos. Second, you can take videos of planets using either a webcam or DSLR with "crop mode" functionality. In my opinion, the main difference is that planetary photography is much more difficult!

You can take single shots of planets at high magnifications using the eyepiece projection method. If you do, you are going to want to enable some sort of "mirror lockup" to prevent blurry images.

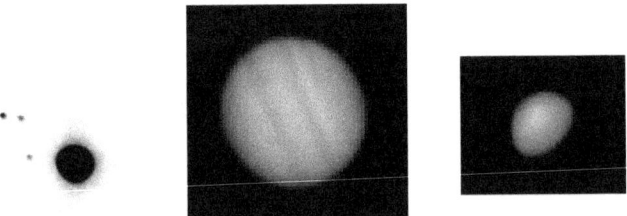

Left: Brighter images reveal moons (Jupiter). Single Shot.

Center: Darker images reveal planetary detail (Jupiter). 1200mm focal length. 90mm refractor. 6 second DSLR video @60fps. Best 70% stacked.

Right: Venus. 1200mm focal length. 90mm refractor. 6 second DSLR video @60fps. Best 70% stacked.

DSLR Astrophotography (DSOs)

When it comes to DSOs, you are going to need an equatorial mount and a telescope. No way around this one. Refer to the chapters on buying a telescope and equatorial mounts for information related to these items.

Before you proceed, I am sure some people are thinking, "why not just use a longer focal length camera lens?" Great question. The answer? Telescopes gather light. Use a telescope! The difference between f/5 (common telescopes) and f/5.6 (common camera lenses) is noticeable in terms of image brightness. Telescopes are also designed to be focused to infinity, so that is where their optics are the sharpest. This is very good for us.

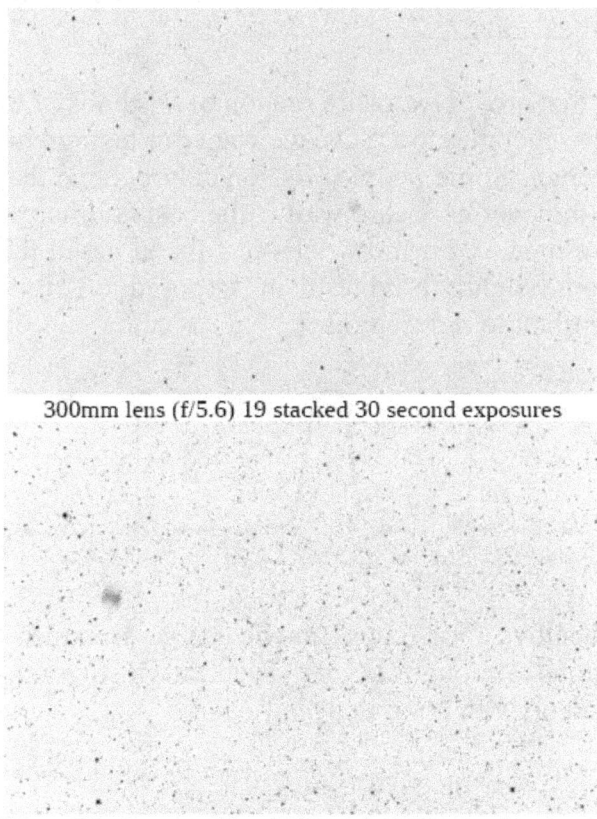

300mm lens (f/5.6) 19 stacked 30 second exposures

400mm f/5 telescope (80mm) 20 stacked 20 second exposures

We are only going to use the prime focus method in this book. As a reminder…

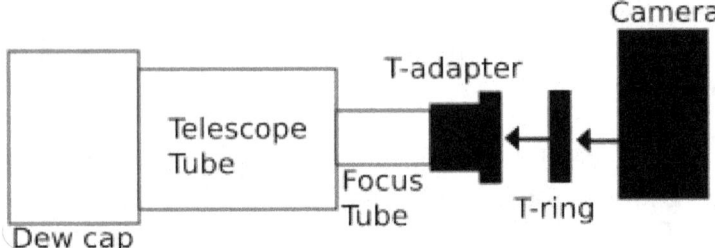

When imaging DSOs under light polluted skies, you may wish to use a UHC or LPR filter between the T-adapter and focus tube. The filter should screw into the T-adapter. When you use such a filter, you will need to use longer shutter times because the resultant image will be darker.

Remember that proper exposure section of this book? Let's go back to that. When it comes to DSOs, we want our histogram to *at least* be in the center. Some people passionately believe that ¾ of the distance to the over-exposure wall is the best (similar to "exposed right" in the image). For now, let's just try to get in the center of the histogram (slightly brighter than "exposed left" in the image). This is important to reduce noise!

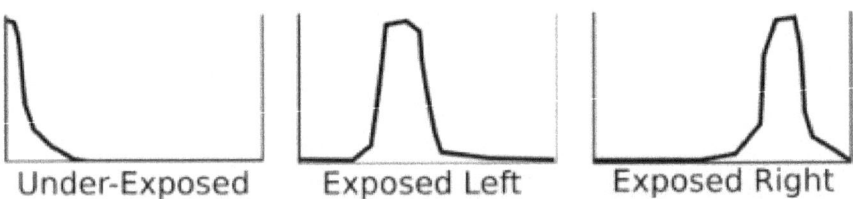

If you can achieve good focus (maybe using a Bahtinov mask) and track accurately enough to get a decent histogram, you will probably end up with good images.

We will discuss stacking images later, but a common question that comes up is...

Is it better to stack a bunch of short exposures or stack a few longer exposures?

Unfortunately, the answer usually disappoints. When it comes to DSOs, we need a good histogram and a lot of collected data. The only way to do this is through long exposures with sufficient tracking. Let's get a little more into this.

No tracking method: Obviously, this is the cheapest way. All you need to do is mount your camera or telescope on a tripod and take the longest exposures you can before you see stars trailing. You will need to use the 500 or 300 rule. I should remind you that every photon counts!

The images on the next page are both of M4 and show the difference between stacking 25 and 200 short exposures. The difference is much more dramatic in non-inverted colors!

Equatorial mount: You will need a right ascension (RA) motor. We will discuss this more in the chapter on equatorial mounts.

Most photographs of DSOs in this book were taken using an equatorial mount with an RA motor. There is really no other way to do it!

25 stacked 1.6 second exposures. Barely visible!

200 stacked 1.6 second exposures. Ahh, there it is!

When there are clouds, go inside. When the moon is up, direct your telescope towards open clusters and globular clusters (or do planetary and lunar). Nebula and galaxies should only be photographed on moonless, clear nights. This perfect combination happens a lot less often than you would like!

Dedicated Astronomy Cameras (DSOs)

Honestly, I got sick of using DSLRs relatively quickly before I upgraded to dedicated astronomy cameras. Sometimes, you must use the right tool for the job.

Dedicated astronomy cameras are specially engineered for DSO and/or planetary/lunar imaging depending on the make and model.

DSO models should be cooled to reduce thermal noise. This makes them a huge upgrade over DSLRs. There is a lot that goes into these things, so we'll summarize by saying, "they're just better".

Monochrome (Mono) or one-shot-color (OCS)?

In my opinion, there is no question about the answer to the DSLR vs dedicated astronomy camera question. Dedicated astronomy cameras are always better, but there are *plenty* of talented people in the world who showcase the power of DSLRs in astrophotography. I am not one of them. Regardless, dedicated astronomy cameras are always better because they were designed for this purpose.

The real question is Mono or OCS? OCS cameras are a lot like a DSLR. The photo you take has color. Mono cameras take images with no color. Why would you want that?

OCS cameras (both dedicated astronomy OCS & DSLRs) have a Bayer filter that essentially contains 25% red, 25% blue, and 50% green pixels.

Monochrome cameras have no such filter. They suck in all the light they can, regardless of color. They are very effective and sensitive.

In order to create colored images, you need to take filtered images and combine them into a final photograph. For example, you would need to take 25 images with a red filter attached to your camera one night, take 25 images with a green filter attached to your camera another night, and take 25 images with a blue filter attached to your camera another night and combine all 3 sets of images in order to create a RGB image. This is a lot of work, but this is how high-quality work is usually produced.

Monochrome sensors are even better if you wish to shoot narrowband images of emission nebula.

Most emission nebula release photons with wavelengths of 656.3nm (hydrogen alpha) and 500.7nm (oxygen-III). You can use filters that only pass these wavelengths and collect that pure emission data. It is very nice, even in heavily light-polluted skies. It makes astrophotography possible in cities.

OCS cameras absolutely struggle here. Why? Because of that Bayer filter. If you attach a hydrogen-alpha filter on your OCS camera, only 25% of the pixels can even register the photons. That is NOT good!

100% of the pixels on a monochrome sensor can register the same photons. If you want to or must shoot narrowband, you need to get a monochrome dedicated astronomy camera with at least hydrogen alpha and oxygen-III filters. Now you can take 25 images with the hydrogen-alpha (H-A) filter attached one night and take 25 images with the oxygen-III (O-III) filter attached another night and combine the two images into a RGB image assigning the H-A data to the R channel and the O-III data to the G and B channels.

You can combine different wavelengths into RGB images in many ways, but I find the H-O-O method to the cheapest and easiest way.

In order to pull this style of astrophotography off, we need to expose our images for at least several minutes. With my 50mm f/5 telescope and monochrome camera, I like to take 10-minute exposures. In order to achieve this, we really need good tracking & autoguiding (chapter 15). There is no way around it.

Clearly, I am a fan of collecting H-A and O-III data and combing them into an RGB image. It's easy to do even from a light-polluted backyard. The processing side becomes a little more difficult, but if you enjoy developing computer skills, it should be an enjoyable learning experience.

The type of H-A and O-III filters you use is another source of great debate in astrophotography groups and online forums.

The bandwidth of a filter refers to the range of wavelengths a filter transmits to your camera's sensor. A 7nm filters passes more light to your camera's sensor than a 3nm filter does. Note that both filters pass the same amount of signal to the camera's sensor. The wider filter simply passes other light sources too, but 7nm is a good compromise between cost and effectiveness.

My opinions…

H-A: Get a 5-7nm filter, unless you want to dedicate your work to collecting data on planetary nebula. The N-II emission line at 658.4nm is included with 5-7nm H-A filters but excluded from 3nm H-A filters. For emission nebula work, it is nice to get the boost in signal from the N-II line included with the H-A filter, so the 5-7nm filter is nice. For planetary nebula work, you may wish to separate N-II and H-A into different color channels, so the 3nm filter becomes necessary.

O-III: Get a 3-7nm filter. O-III is generally more inclusive of moonlight and light pollution than H-A. I can collect H-A data with a 7nm filter during a full moon and actually collect good data. O-III data is best collected with a narrower filter and during moonless nights, but 7nm should still be adequate if the sky is moonless.

If you get sick of the bicolor HOO style, you can always add sulfur (Sii) to your filter collection. This allows you to add a bit more depth to your photos. The SHO color scheme is a very popular one (S:R, H:G, O:B).

The case for OSC cameras

While I stated earlier that mono is best, OSC is making a comeback in recent years (2021 update). I actually just purchased an OSC camera. What changed?

First off, the quantum efficiency of cameras is getting so good these days, that a Bayer matrix doesn't bother people as much as it used to. Just expose for longer, who cares?

Second, the development of multi bandpass narrowband filters has made OSC a better idea. These multi bandpass filters allow, for example, the pass of Ha and Oiii through the filter at the same time. This means that your OSC camera can now collect R, G, and B narrowband data at the same time. That is a huge time saver for many busy people, and it makes processing the data that much easier.

Unless this is your full-time job, astrophotography can be quite a demanding hobby. It requires late nights on work nights and sometimes more than 5 of those nights a month is just not possible. OSC imaging makes sure you are collecting colorful data every night you image. It also allows for a simpler setup, which saves time. No more worrying about filter wheels and $3000 worth of narrowband filters laying around. OSC is simpler and sometimes there is a great deal of value in that.

It is still argued that mono is better, and while I agree completely with that statement, my life has just gotten too busy to worry about all those things. I look forward to my journey with my new OSC!

How to buy a dedicated astronomy camera for DSO use

1. Color or Mono?
   a. Only you can answer this question.
2. Cooling
   a. Cooling is needed.
3. Diagonal
   a. You need to determine the image circle of your telescope or field flattener / coma corrector. Thanks to wonderful marketing strategies, you may want to use a sensor with a sensor diagonal smaller than the listed image circle by a good margin.
4. Pixel Size
   a. You should aim for an image scale of 1-2"/pixel. Anything smaller than 1"/pixel is going to require significant effort, but it is still very possible. It depends on the style of astrophotography you are interested in. An image scale of 1.5"/pixel is still interesting.

$$R \ (''/pixel) \ = \ \frac{pixel \ size \ (um)}{Focal \ Length \ (mm)} * 206.3$$

5. Megapixels (MP)
   a. MPs are slightly over-rated but anything over 20MP will give very high-quality results.
6. Quantum Efficiency (QE)
   a. This is becoming irrelevant now with camera sensor technology getting so good. Still peak QE of 80% is incredible.
7. Dynamic Range
   a. 12 stops is outdated, 14 stops is a good goal.
8. Full Well Capacity
   a. 25ke is good, but 50ke is better. Full well capacity relates to how quickly you clip data by over-exposing bright regions.

Dedicated Astronomy Cameras (Planetary/Lunar)

The only way to get high-quality planetary/lunar imaging done is by taking videos and stacking the individual frames.

Planetary/lunar models do not require cooling. They just need to be able to record video with as many frames per second as possible. The more frames, the better the final image will be.

The monochrome vs OCS debates comes up once again. Clearly, when it comes to DSOs, I believe that the camera should be monochrome unless you are lucky enough to live away from light pollution. When it comes to planetary, I am not experienced enough with both styles to give you advice either way.

I believe the answer depends on your telescope and location.

My location is terribly affected by light pollution and unsteady atmospheric conditions. Really terrible for astrophotography. For DSO work, I get around this by using narrowband filters as described in the last section. There is not an easy work-around for planetary imaging. Light pollution doesn't play as big of a role, but the atmospheric conditions certainly do.

If you go monochrome, you will need to get a filter wheel with R, G, and B filters. No way around it if you want colored images.

Things get complicated here because not only do we need some sort of tracking to compensate for Earth's rotation, we must worry about the rotation of the planet we are imaging too.

We could talk about the mathematics here for several pages, but let's take a break and remember the trial-and-error method. Get practice working with videos ~3 minutes long. If you start to see planetary rotation blur the image, shorten the video. There is a lot of software out there that can handle planetary de-rotation. Learn about that on your own.

If you go monochrome, you will have to take your R, G, and B videos within that 3-minute timeframe. That is why you need a filter wheel and preferable parfocal filters, so you do not have to refocus your image. You only have 3 minutes, which is not a lot of time. OCS is easier here, because you can simply record a single 3-minute video and process it relatively quickly.

Me? I use a monochrome sensor. That seems confusing since I just ended the last paragraph highlighting the simplicity of an OCS sensor for lunar and planetary applications. It all depends on your goals.

I have a 12" dobsonian that I put two motors on (up-down & left-right motion) so that I can track planets and the moon as the Earth rotates. Large telescopes are more affected by atmospheric turbulence. Why? Don't over-think this one. The mirror on a 12" telescope is much larger than the objective lens on a 4" refractor. The 12" mirror simply has to look through more turbulence than the 4" refractor.

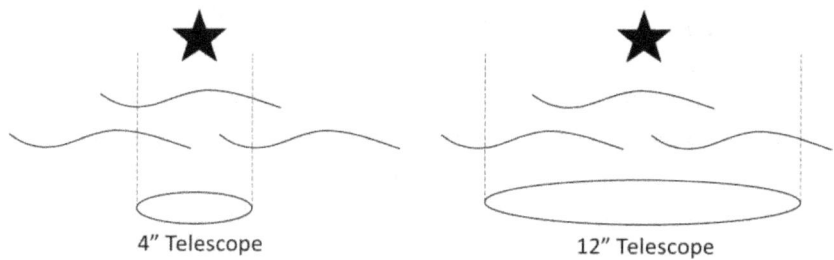

4" Telescope   12" Telescope

Longer wavelengths of light are less affected by atmospheric turbulence. IR light has longer wavelengths than visible light. So, an option for black-and-white, colorless planetary and lunar imaging for large telescopes that is less affected by poor seeing conditions is IR-only lunar and planetary imaging. IR-pass filters are available that only transmit IR light to the camera's sensor. I currently use a 742nm IR pass filter. The resultant image is darker and has no color but is generally clearer. While not an ideal method, it is one method for large scopes that avoids a filter wheel purchase.

But like all other things in this hobby, it is a compromise.

Longer wavelengths are less effected by seeing, but you get less resolution out of your telescope! I know, it's ridiculous.

Great seeing? Want the highest resolution possible? A lot of people use a green filter on their monochrome sensors for high-resolution photography.

Bad seeing? Want to get anything decent out of the night? A lot of people switch to IR pass filters on their monochrome sensors for decent imaging during subpar conditions.

Shorter wavelengths: Better resolution, but more effected by seeing

Longer wavelengths: Worse resolution, but less effected by seeing.

A discussion on how to buy a dedicated astronomy camera for planetary & lunar work can be found in chapter 13.

Dedicated astronomy camera lunar gallery

203mm SCT

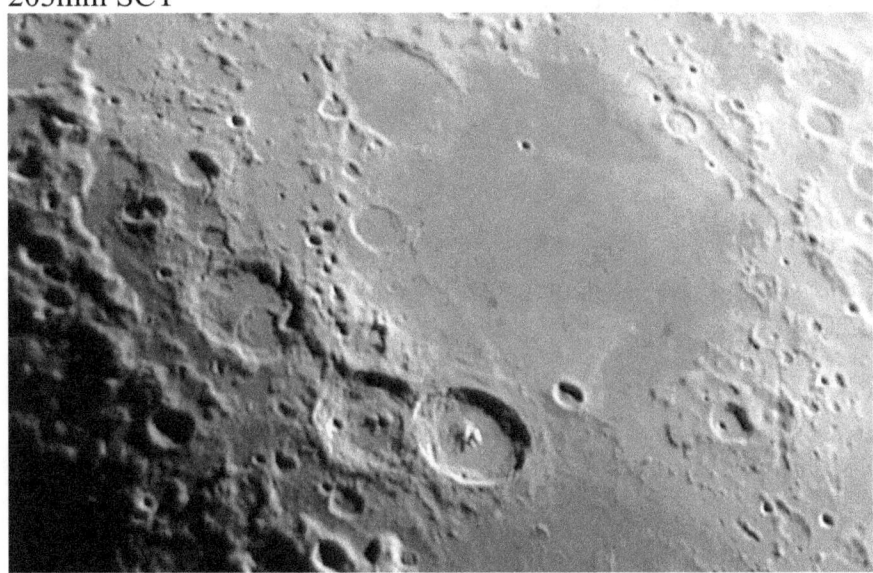

203mm SCT (mosaic)

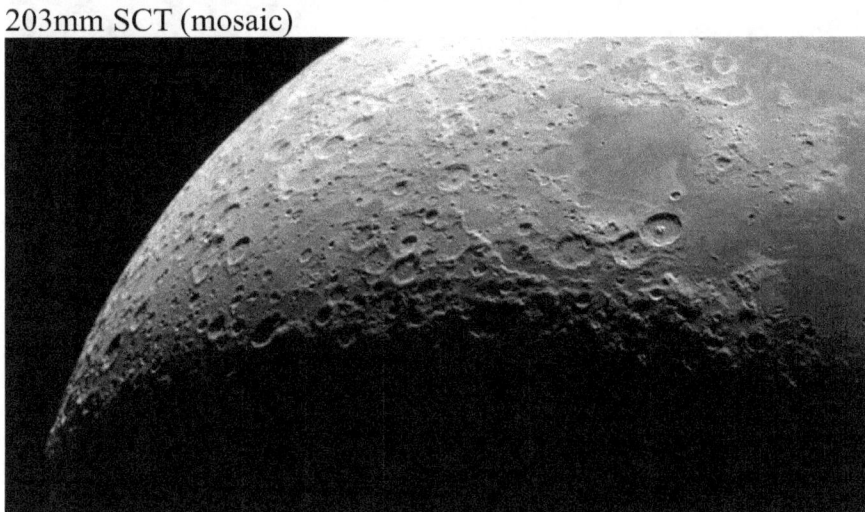

203mm SCT

12" Dobsonian

DIY 6" f/8 Reflector

254mm SCT

Chapter 11

Finding Astrophotography Targets

The magnitudes of these objects are in parenthesis. Most of these are challenging targets for small telescopes.

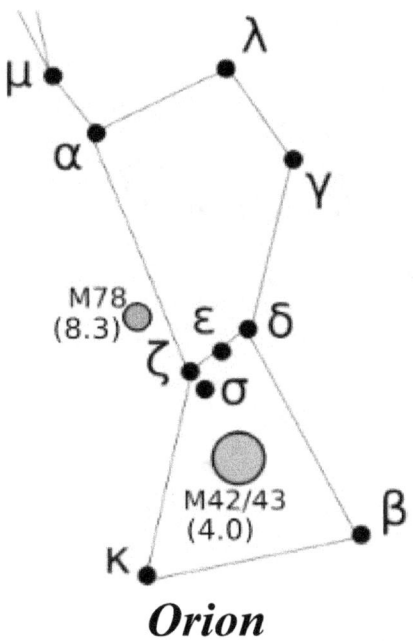

Orion

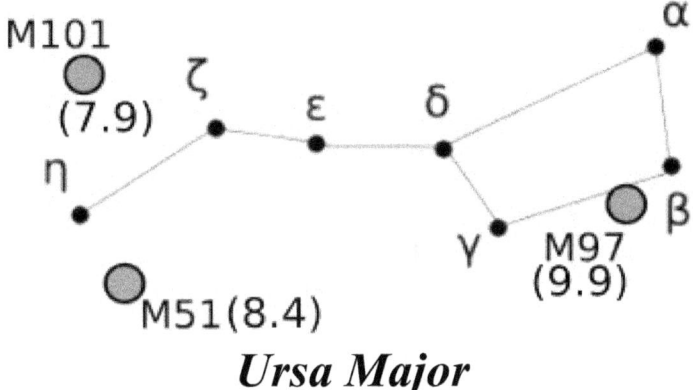

Ursa Major

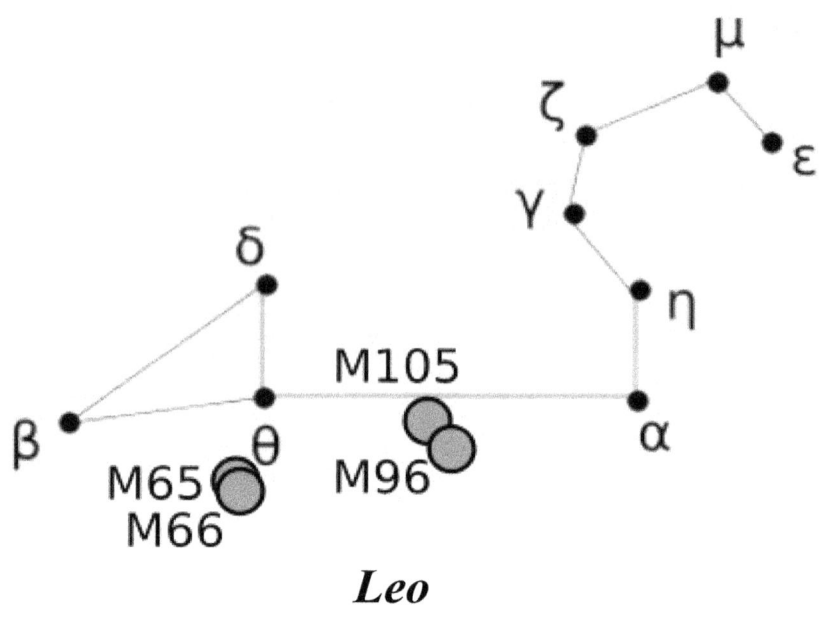

Leo

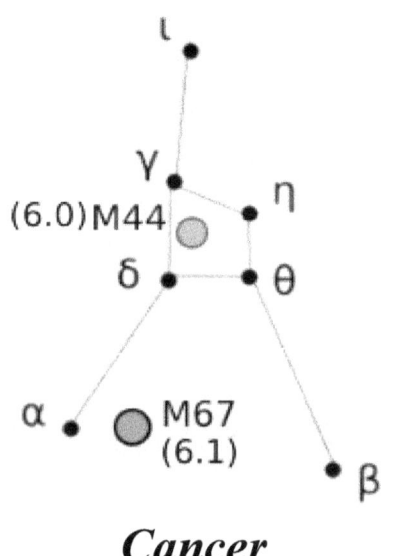

Cancer

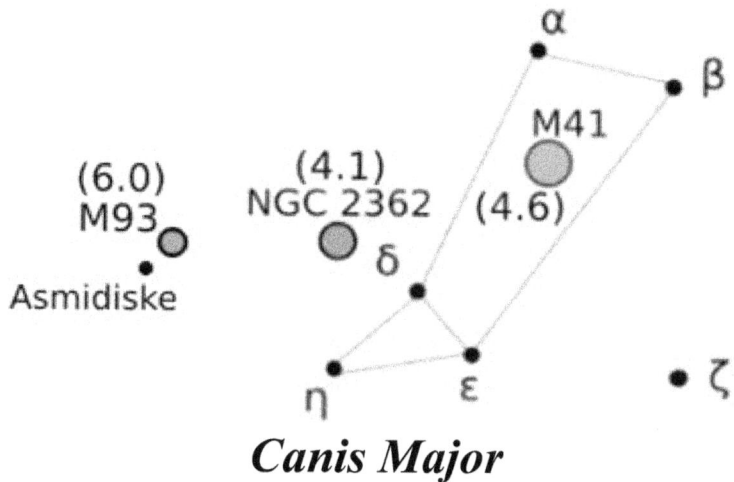

Canis Major

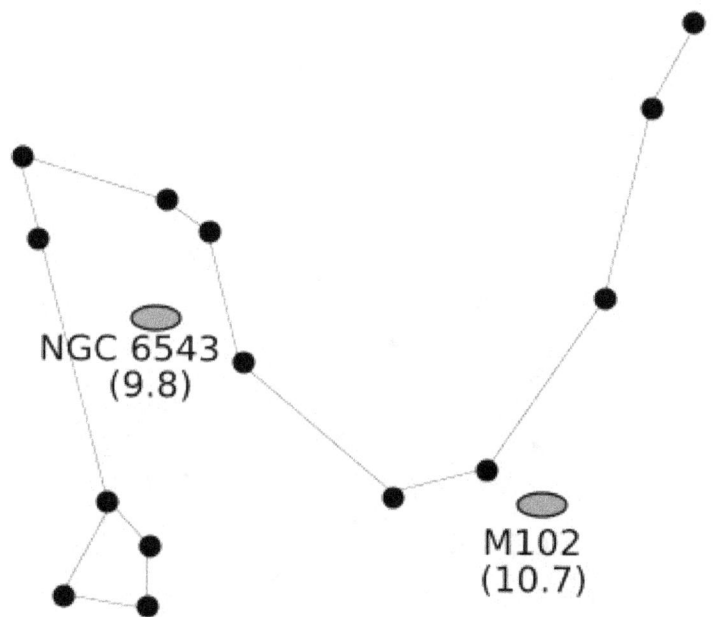

Draco

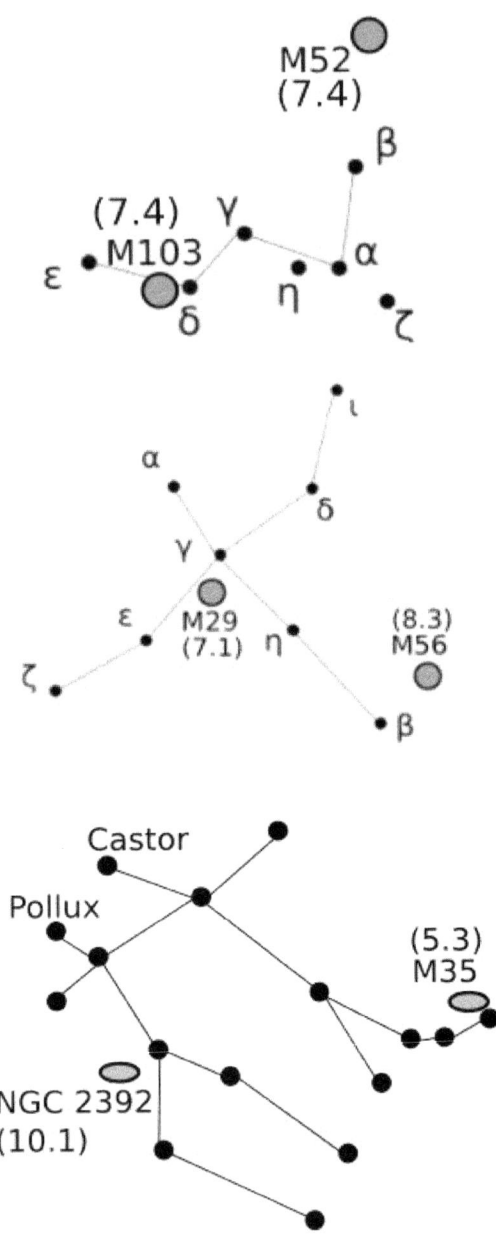

Cassiopeia (top), Cygnus (middle), Gemini (bottom)

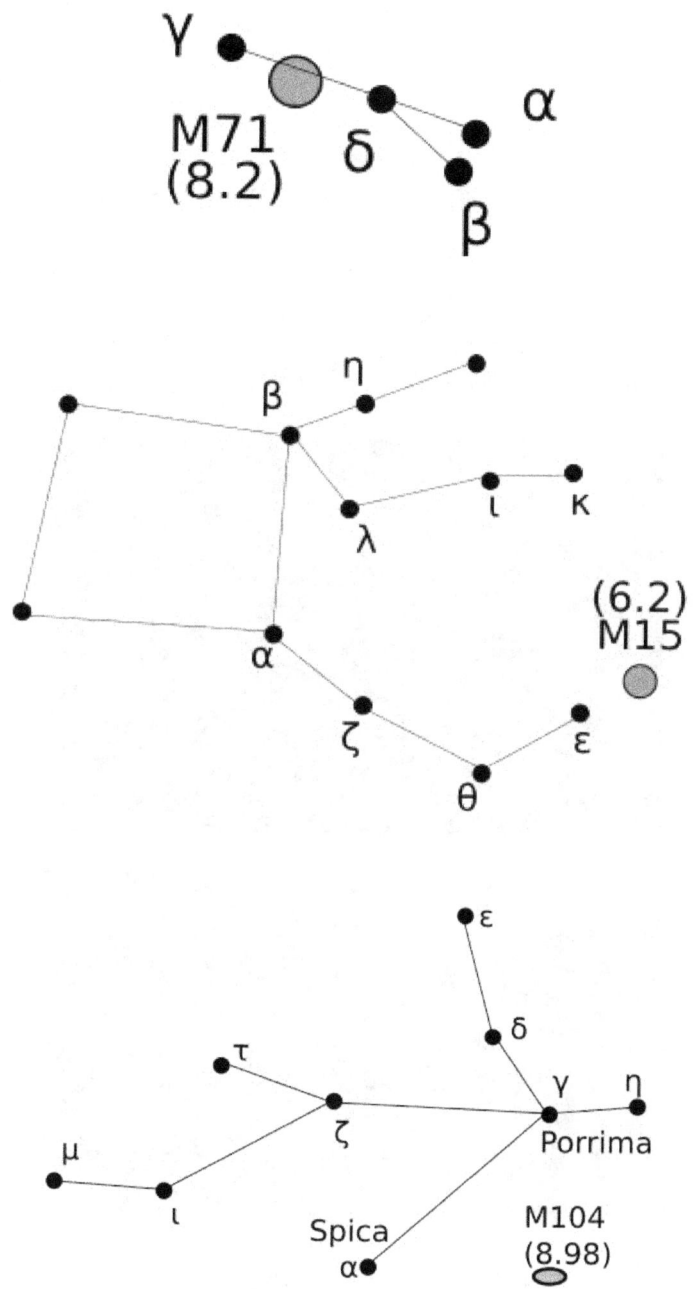

Sagitta (top), Pegasus (middle), Virgo (bottom)

What are the best DSOs to image? This is a very well-debated topic, but I can provide you a list of some of my favorite beginner targets.

M4, M13, M8, M28, M31, M42 and M45 are easy.
Leo triplet, M57, M27, M51, and M1 may be more challenging.

M4: *Bright & Easy.* M4 is a fantastic globular cluster. A 300mm lens can even handle this target. Just point your telescope or camera towards Antares and Alniyat. You should be able to see both of these stars through the viewfinder of the camera.

If you need a good first astrophotography target and M4 is visible, I suggest you attempt it.

M13: *Bright & Relatively Easy.* Find the square in the center of the Hercules constellation and look for HIP 81848. M13 is very close to that star.

Eta Hercules

M13

HIP 81848

Zeta Hercules

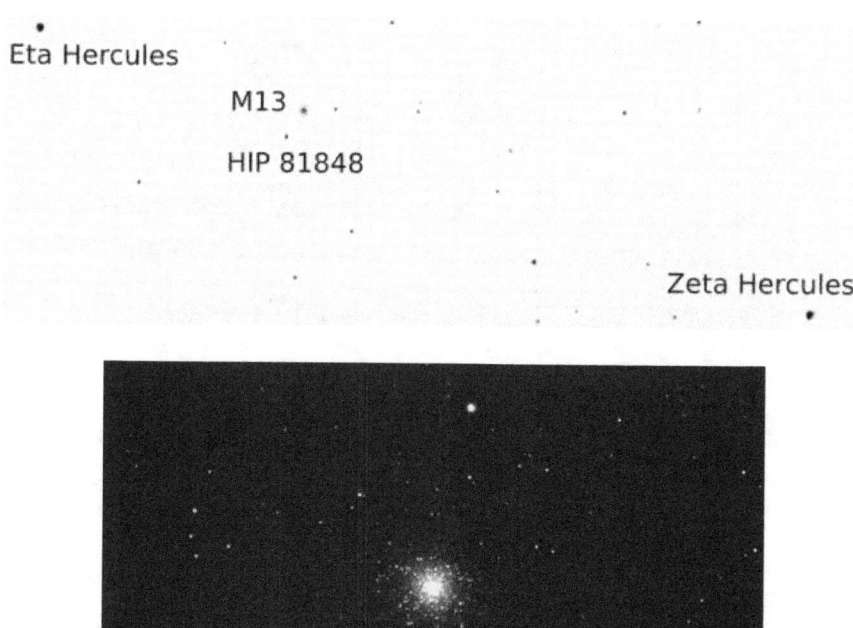

M13 is a fantastic summertime object. Do M4 first though!

A note on globular clusters:
Globular clusters simply contain stars, so you are able to photograph these objects during a full moon. The same is true for open clusters.

M8: *Big, Bright, and Easy.* You can fit M8, M20, and M21 all in the same frame using a 300mm lens. The 11 Sagittarius, 4 Sagittarius, and HR 6717 triangle is fairly easy to find.

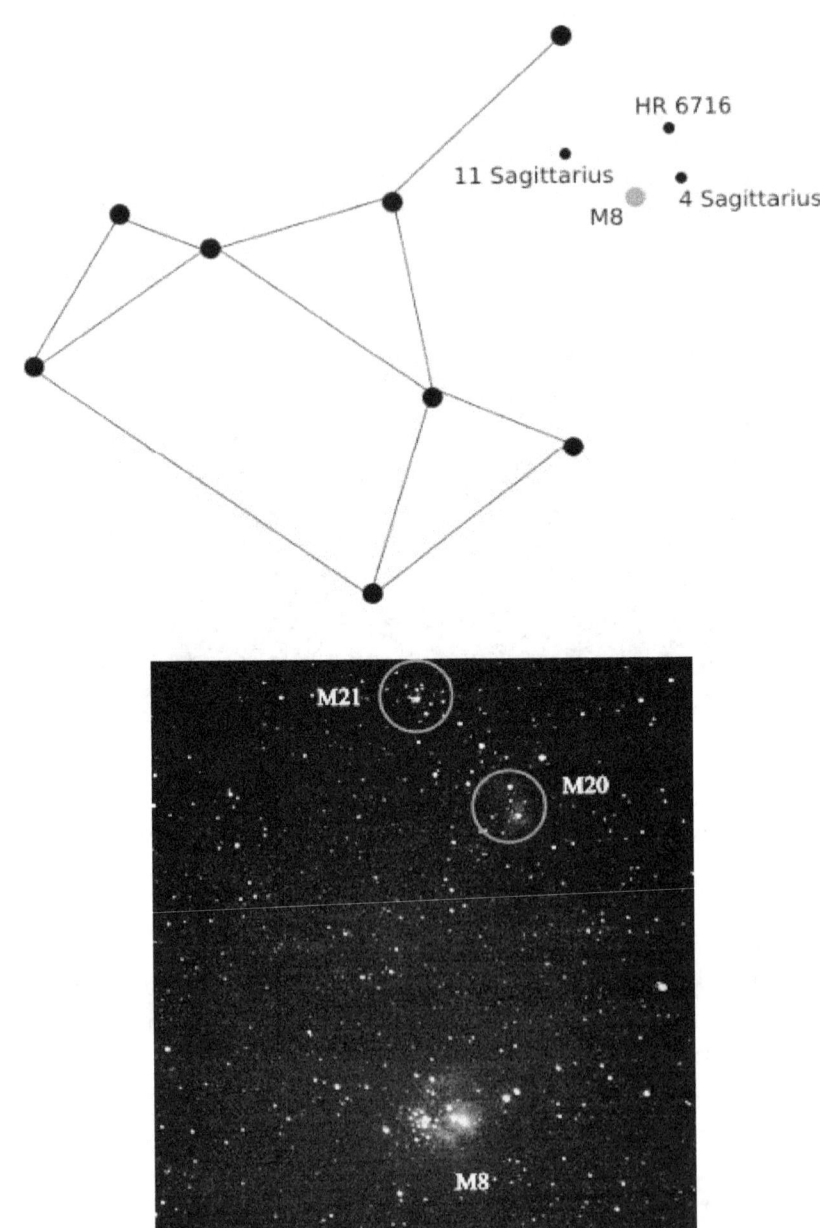

There are many summertime Sagittarius DSOs worthy of observation and astrophotography. The additional stars shown in the image below might help make them easier to find.

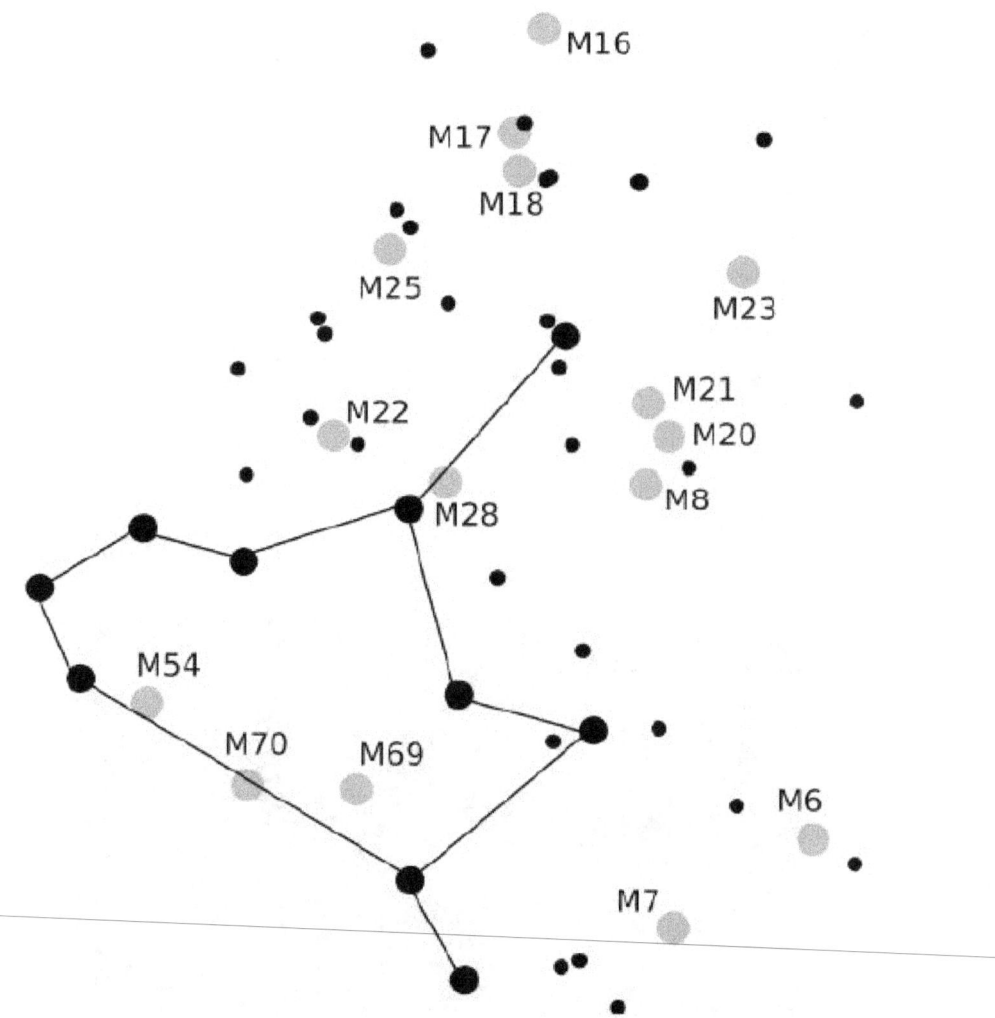

M28 is particularly easy to find and that's why I suggest it. M17 is also a great target to try once you get a little experience.

M31: *Big, Bright, and Easy.* M33 and M34 are also nearby. Everyone loves M31 so it is worth an attempt!

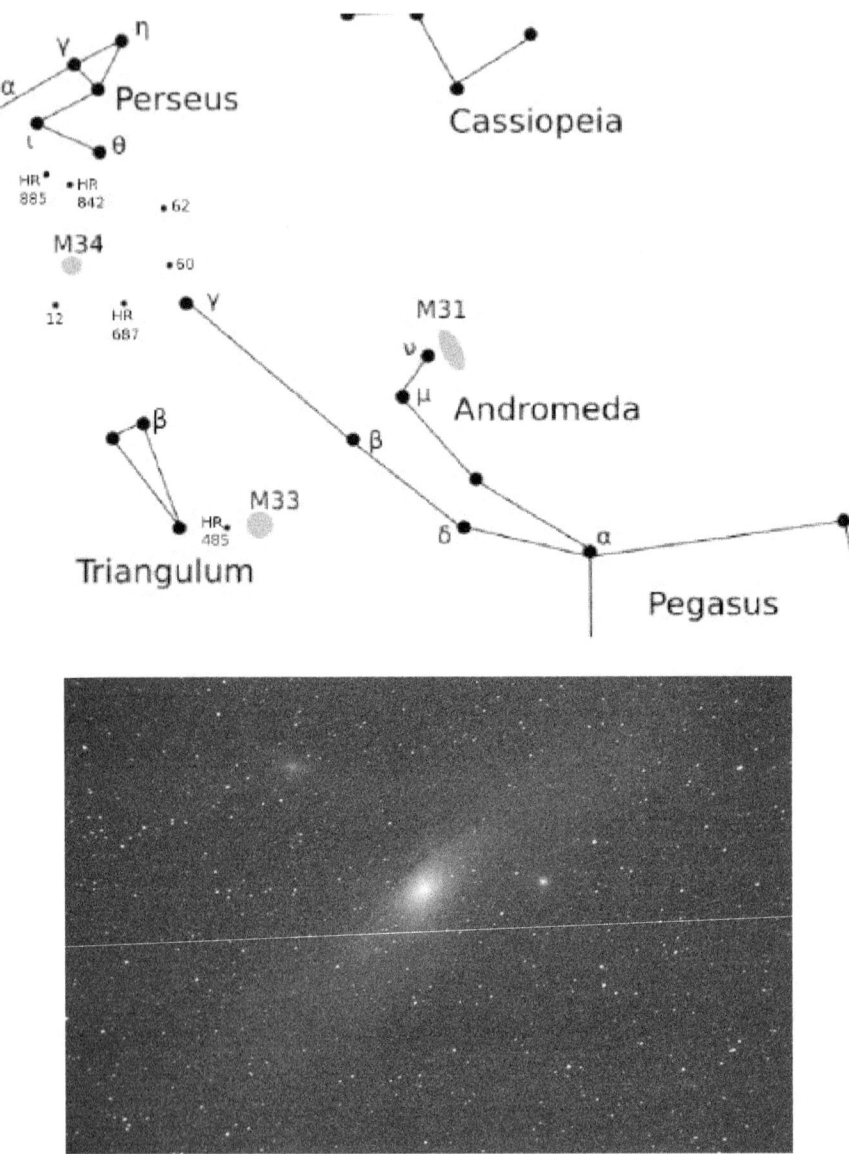

M42: *Big, Bright, and Easy.* Just point to those stars under Orion's belt! This image is only 40 stacked 20s exposures.

M45: *Big, Bright, and Easy.* This is a very popular astrophotography target. You should be able to find it without a telescope.

M4, M13, M8, M28, M31, M42, and M45 are my favorite beginner-level targets. Once you successfully get these on camera, you are ready for some better challenges!

The double cluster is another "easy" target. I took this image during a full moon.

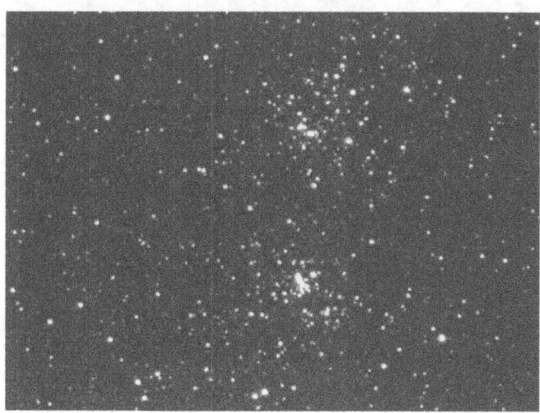

Leo Triplet: *A reasonable challenge.* This target is going to frustrate you! In my opinion, the Leo triplet is relatively difficult for a beginner to find. The three galaxies will likely show up very faint, but long exposures should reveal them.

M57: *Easy to find at <600mm focal length.* M57, the Ring Nebula, is a very interesting target. If my 80mm refractor can reveal the ring with just a single 20s exposure, you can too! Look between gamma Lyrae (left) and beta Lyrae (right) (a double star).

M27: *Achievable, but difficult.* I look for something I call the "Vulpecula House". You could also trace a line that starts at Sigma Sagitta and goes through Zeta Sagitta. M27 is very faint so it may be easier to take quick 5 second exposures until you see it.

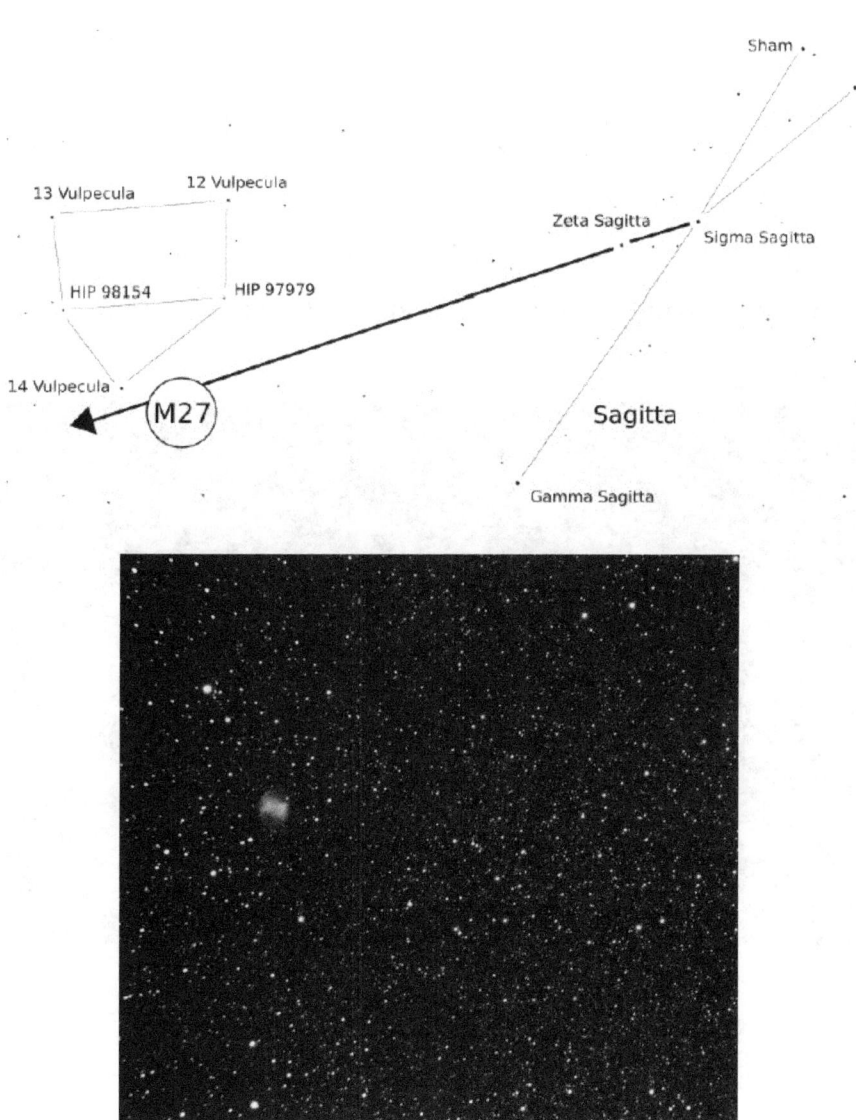

M51: *My favorite challenge.* Find the Big Dipper and find Benetnash. M51 form a sort of right triangle with Benetnash and Mizar.

24 Canes Venatici, HIP 66004, and HIP 65765 are good stars to find.

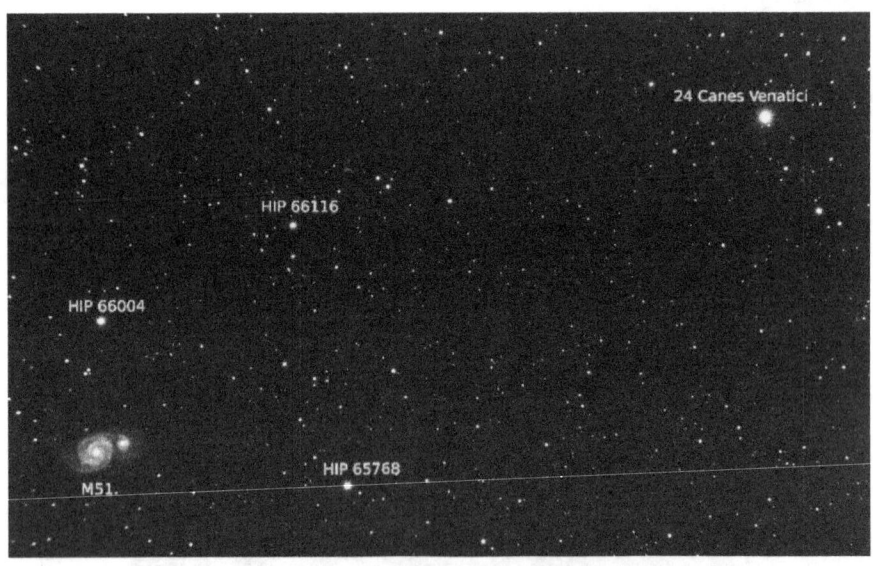

M1: *Easy to find.* Despite its easy location, M1 may be difficult to reveal.

Want more? Below is a table listing the Messier objects with a magnitude less than 6.

Object	NGC	Magnitude	Size (arc-min)
M45	0000	1.6	110
M31	224	3.4	178x63
M44	2632	3.7	95
M42	1976	4.0	85x60
M7	6475	4.1	80
M6	6405	4.2	25
M24	6603	4.6	90
M41	2287	4.6	38
M39	7092	4.6	32
M22	66656	5.1	24
M47	2422	5.2	30
M35	2168	5.3	28
M34	1039	5.5	35
M48	2548	5.5	54
M4	6121	5.6	26
M5	5904	5.6	17
M33	598	5.7	73x45
M13	6205	5.8	17

More details about finding M2, M39, M3, and M5 can be found on the following pages.

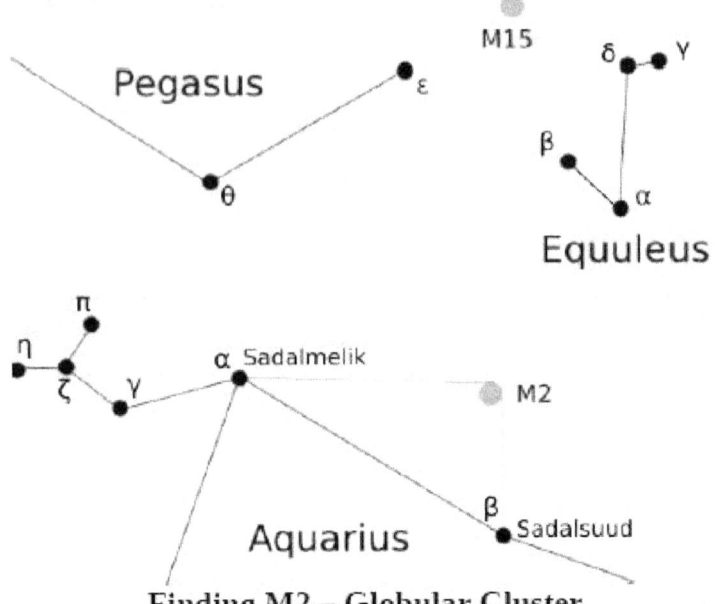

Finding M2 – Globular Cluster

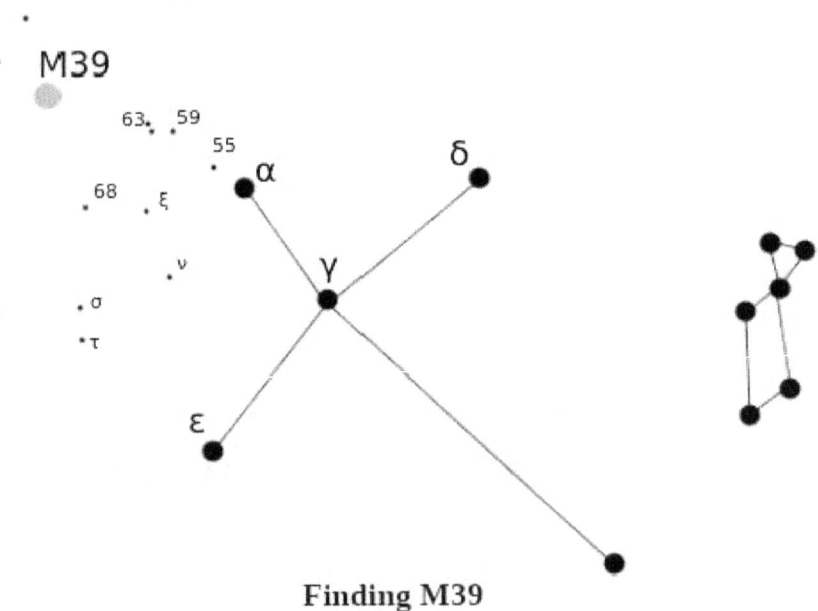

Finding M39

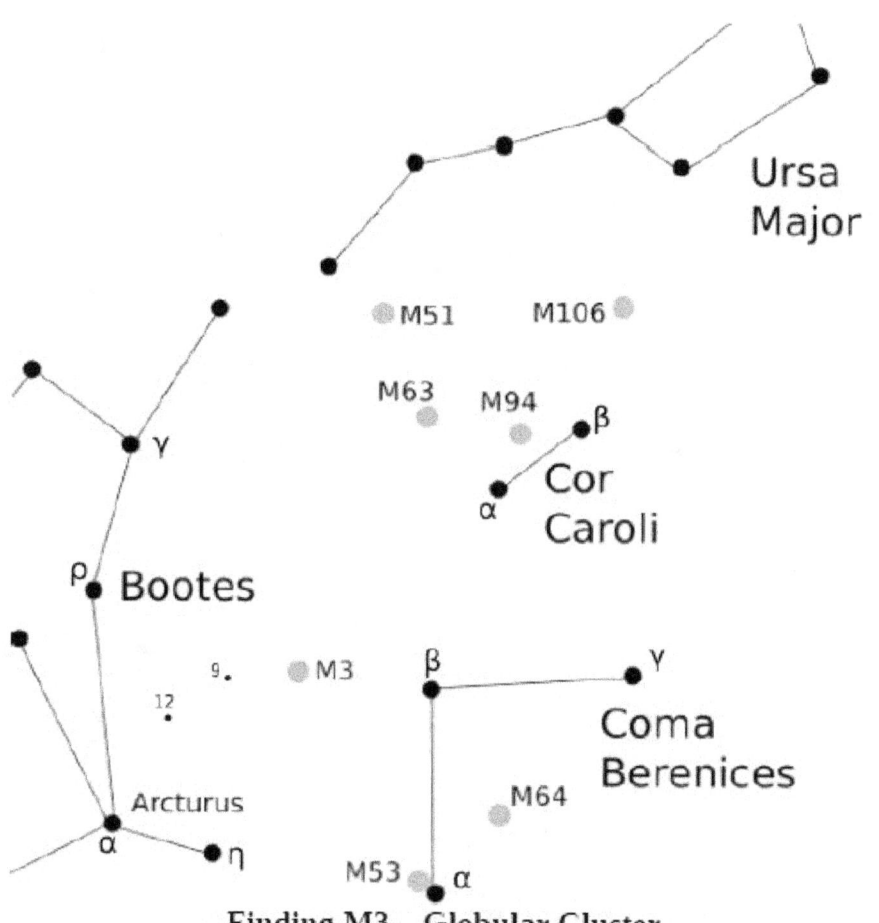

Finding M3 – Globular Cluster

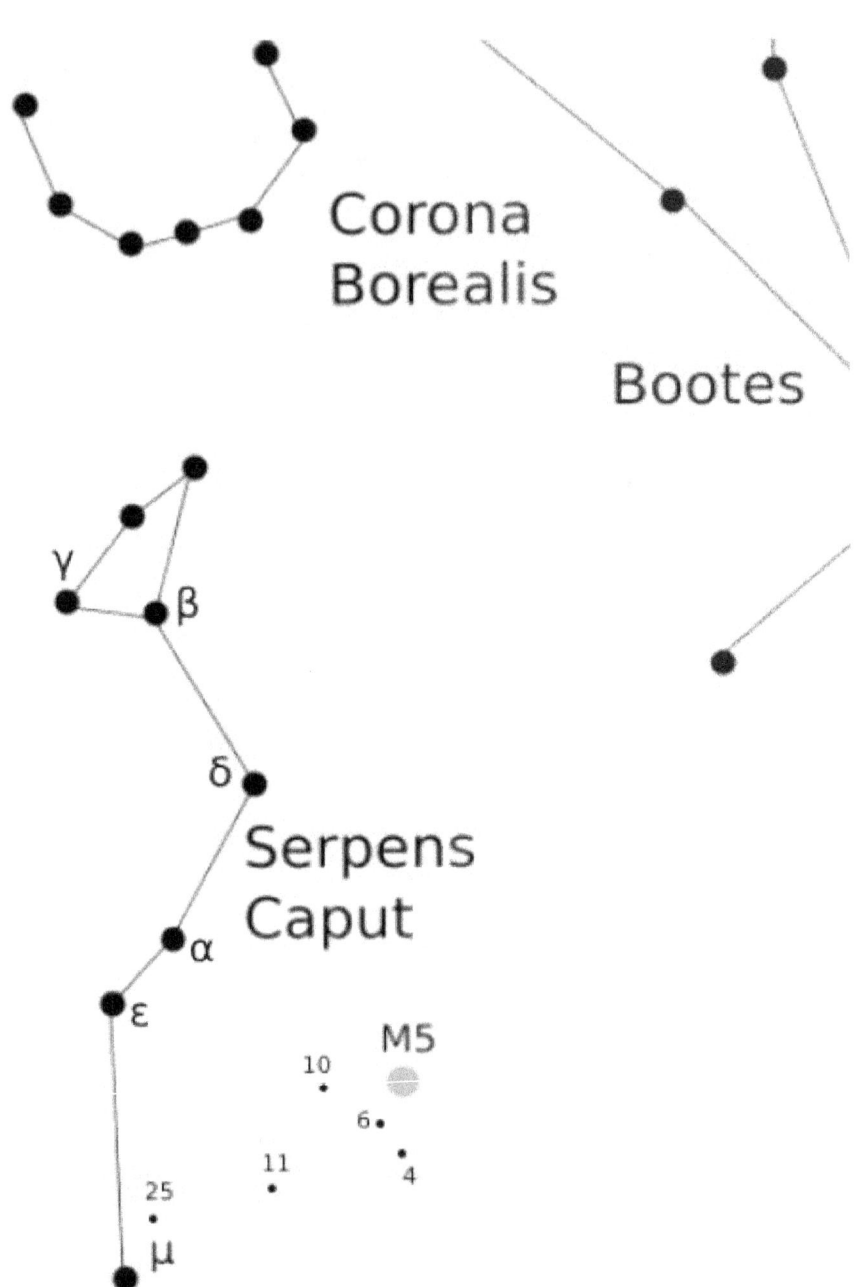

Corona
Borealis

Bootes

γ
β

δ

Serpens
Caput

α

ε

M5

10

6

11
25
4

μ

Finding M5 – Globular Cluster

Galaxies with lower magnitudes are not always the brightest. If this seems confusing, trust me, you are not alone! When you think about imaging a galaxy, you have to consider surface brightness. I have attached a table for your convenience. I have tried to include the more popular galaxies you may wish to pursue.

Galaxy	Surface Brightness
M31	13.3
M33	14.1
M32	12.0
M51	13.0
M63	13.3
M81	13.0
M82	12.4
M83	12.5
M101	14.6
M49	12.7
M74	13.9
M106	13.4
M110	13.3

A "funny" example of this happened to me. At the time, I had successfully photographed M51. M51 has a magnitude of 8.4, so I decided to shoot M101 because it has a magnitude of around 7.8. Since the magnitude was lower, I thought it would be easy! I was wrong and discovered that M101 is a difficult target due to its low surface brightness.

Narrowband Imaging Targets

Below is a list of narrowband targets if you decide to go down this path. It is a great beginner list. Some examples are shown on the next two pages. All of these were taken using a 50mm f/5 telescope and a dedicated monochrome astronomy camera.

Season	Prefix	Number	Name	Size (X')
Summer	M	8	Lagoon	90
Summer	M	16	Eagle	35
Summer	M	17	Omega	46
Summer	M	27	Dumbbell	15
Summer	NGC	6888	Crescent	20
Summer	IC	5070	Pelican	80
Summer	NGC	7000	North American	150+
Summer	NGC	6960	Western Veil	70
Summer	NGC	6992	Eastern Veil	60
Summer	IC	1396	Elephant Trunk	150
Summer	NGC	7293	Helix	20
Fall	NGC	7380	Wizard	30
Fall	Sh2	155	Cave	60
Fall	NGC	7635	Bubble	40
Fall	NGC	281	Pacman	40
Fall	IC	1805	Heart	150
Fall	IC	1848	Soul	120
Winter	NGC	1499	California	145
Winter	IC	405	Flaming Star	30
Winter	M	42	Orion	66
Winter	IC	434	Horsehead	60
Winter	Sh2	261	Lower's	45
Winter	NGC	2174	Monkey Head	40
Spring	IC	443	Jellyfish	50
Spring	NGC	2244	Rosette	80
Spring	NGC	2264	Cone	80+
Spring	IC	2177	Seagull	120

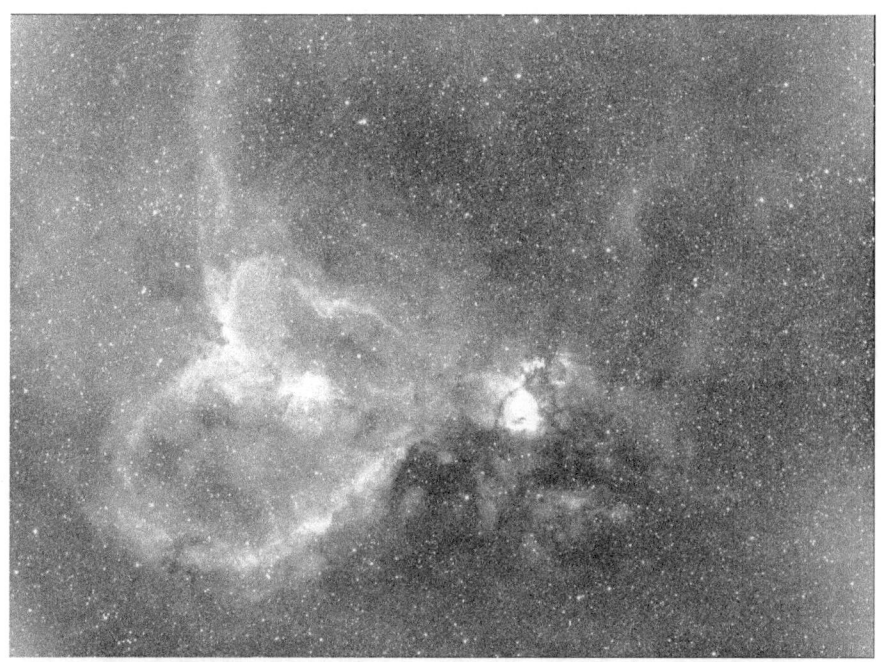

The Heart Nebula (3 hours H-A data)

Horsehead Nebula (6 hours H-A data)

The California Nebula (6 hours of H-A data)

Chapter 12

Editing Astrophotos

Post-processing in astrophotography is quite a complex subject. There are a lot of different programs out there and most of them are very confusing if you have never looked at a photo editor before. I use the free program "GIMP" but all photo editors are very similar to each other. The procedures I discuss should be similar.

Lunar & Planetary: After you process your video, you can upload it into a specialized lunar and planetary editor or a photo editor. You can also do the same for a single-shot.

A specialized lunar & planetary editor will have some sort of "wavelet" editing that is very useful. Small adjustments make a big difference. Other helpful edits are "contrast", "brightness", and "gamma". Overall, this editing is relatively straightforward until you become skilled enough to learn more. The software you choose to use will have its own suggestions and instructions.

DSOs: This is where editing gets a little crazy. In fact, almost every DSO requires its own editing procedures. To make matters even worse, the same DSO can require different procedures if the photos were taken under even slightly different conditions.

After you get your accurately-tracked, neutrally exposed photographs, you will need to "stack" them somehow. There are some good free programs available for stacking DSO images. This stacking software is going to ask for a series of files.

Lights are the photos of the DSO you took.

Darks are simple. All you do is place the dew cap cover on your telescope or lens cap of your lens and take photos at the same ISO, temperature, and focus as your lights. What does this really mean? After you are done imaging your object, put the cap on and take 10-20 photos. They should be dark.

Bias/Offsets are also very easy. Keep the cap on and take 10-20 photos at the same ISO and position but at the quickest shutter speed your camera can do (for example, 1/4000 of a second on my camera). Temperature is not that important but they are so easy to take that you might as well re-do them right after you take your darks.

Flats scare a lot of people away. They scared me too until I tried them. They are easy, trust me. Keep the same ISO and focus position. Switch the camera to aperture priority mode "AV". You need to get a bright source of white light. I have found that the easiest way to do this is to use some sort of electronic tablet. Find a way to get the screen completely white. I open up a web browser and go to full screen, but a lot of people open up a text editor. Take the cap off your telescope or lens and place the white screen completely over the telescope or lens. Take 10-20 photos at whatever settings the camera chooses, but make sure to keep the ISO the same. The histogram should be neutrally exposed.

Example:

It is December 12th and it is a crystal clear night. I choose to image M42. I go outside align my equatorial mount and find M42 with my 80mm refractor. I take 50, 20 second exposures at 800 ISO. A plane flew through 6 of them and some sort of wind or tracking error caused 4 bad photos. At the end of the session, I have 40 good photos. Next, I place the cap on my telescope and take 10 photos using 800 ISO, 20s exposures, and the same focus position. Immediately following, I change the shutter speed to 1/4000s and take 10 photos. Finally, I place a white screen over the telescope and take 10 photos at 800 ISO in AV mode. Done!

After you upload the images into a stacking program and run the program, you will get a final image. Do not get excited though, we need to make a lot of edits. Use a photo editor to do this, not the stacking program.

Editing Astrophotos can get really complex, so I am only going to show you the "curves" and "levels" functions. Actually, these two will get you a decent photograph. Once you master these, you can worry about more detailed editing techniques and programs. It is not my place to instruct others on how to use these programs.

Once you upload the stacked image into a photo editor, you are first going to want to open "levels". You will see something like the following image. Of course if your histogram was too dark, you will have no empty space between your data and the left edge (this is bad). In general, we want something like the following.

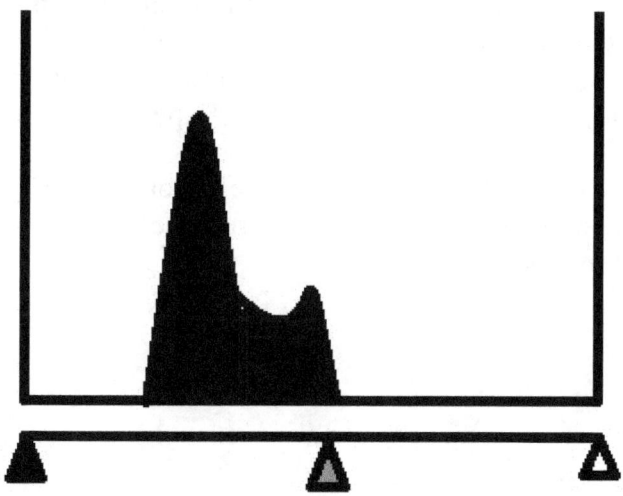

You will notice three "sliders". Ideally, you would want to bring the left-most slider all the way to your data. You can even clip some of the data but this is not recommended at first. Once you move the slider, you will see something like the following.

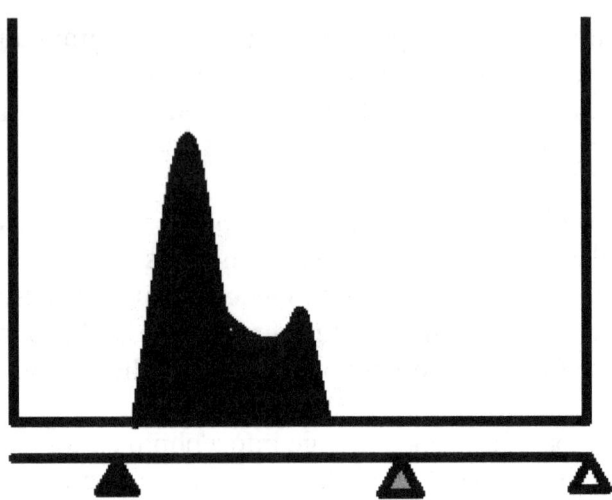

If your image got too noisy, you can go ahead and bring it back over the left. Ideally, this is want we want to do. Now, you can experiment with the right-most slider until you get something you like. Please note that the image should still appear a little too dark before moving on to curves. The right-most slider may blow out stars, so play with the middle slider too. Experiment!

Now, find "curves". Levels and curves should both be under a common category (try "color"). Once you open curves, you should see something like the following,

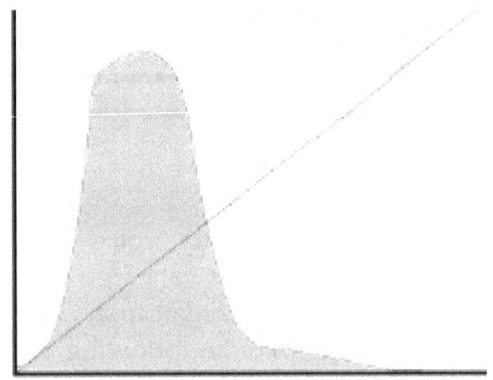

In curves, you can actually glide over pieces of your image and determine where the "good" and "bad" data is. For example, most "good" data (nebulosity) is located around the right edge of the curve and a little passed it on the right side. Most background space data is under the curve.

In order to create a high contrast photo, many people click the right edge intersection of the straight line and the data and bring the right side up (which drops the left side down).

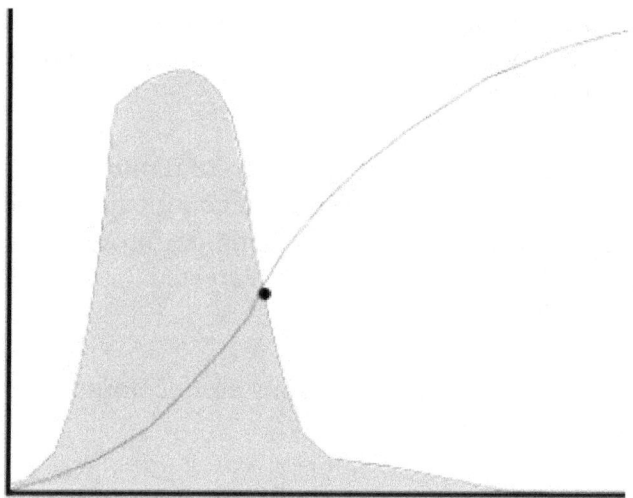

If you do not like this change, experiment with different anchor points and where you adjust the line.

While this is an over-simplified introduction to astrophoto editing, I do believe this is a great place to start and will actually get you decent results after enough practice.

Remember. Good photos are the result of a lot of collected photons. We need long exposures and good histograms!

Pixinsight gets its own page.

Pixinsight is currently the best program available for professional & amateur astrophotographers. There are entire books written about the program. I bought one and find it extremely valuable as I continue to try to learn the program. Be warned: there is a steep learning curve associated with the program. The results are worth the effort.

I built my own computer to run the program as it is very CPU & RAM intensive. Building a computer is actually very easy and there are many online videos and resources showing how to complete a PC build.

A PC build is not required, but it will make running the program a much better experience. It is nice to have a dedicated astronomy computer laying around for all the computer programs you need. This is especially true if you are interested in both DSO and lunar/planetary imaging.

As of 2020, the following components are recommended for a PC build dedicated to astrophotography:

- Computer Case (I went with mini-ITX form factor)
- 8 core CPU is good. More is better. Integrated graphics is enough.
- CPU water cooler – keep your CPU cold. Water-cooling may not be necessary, but I prefer it.
- Motherboard compatible with your CPU
- Int./Ext. HDD (~1TB minimum for videos/photos)
- M.2 SSD (for the operating system & programs)
- 16 GB of memory. More is better, but there are quite a few people who observe the computer rarely utilizing more than 16GB. I went with 32GB to be safe.
- Operating system. I prefer Linux, but Windows has far greater compatibility with astrophotography programs.

Advanced Editing – Single Mono / OSC

In case you are seeking a slightly more helpful guide to astrophoto processing, read this section.

The OSC or single-filter-mono workflow is relatively simple, and the basic structure follows what I presented so far. The only difference with OSC is that you must play a balancing act between the colors. I like to do this in levels and looking at the R, G, and B separately. I can't do this in black & white ink anyway, so I am ignoring it. You'll still get the basics and learn some helpful tips by reading this section.

The same goes for SHO or HOO processing. Since I can't show that in black & white ink, you'll have to learn that on your own. Again, you'll still learn some good strategies by reading through this section.

I will be showing an example of hydrogen alpha data collected using a 73mm f/6 refractor, a 7nm Ha filter, and a monochrome camera.

Your first step is always to acquire good data. You need good data before you can get good photos. You need perfect focus, collimation, thermal equilibrium, and good integration time. For this example, we have 74, 300s images that have been stacked for a total integration time of ~6 hours.

After you get your data, you will stack your photos and calibration files in whatever stacking software you prefer. Note: with the newer dedicated astronomy camera technology out there, you may not need to include darks or offsets. Research what calibration files people are using successfully with your camera to see what is necessary.

After that, load your stacked image into your favorite photo-editing software.

Not very good, huh? Editing is a big part of astrophotography. The first step I take is to adjust the right-most slider in levels and bring that in until the stars start to blow out too much. This is all subjective.

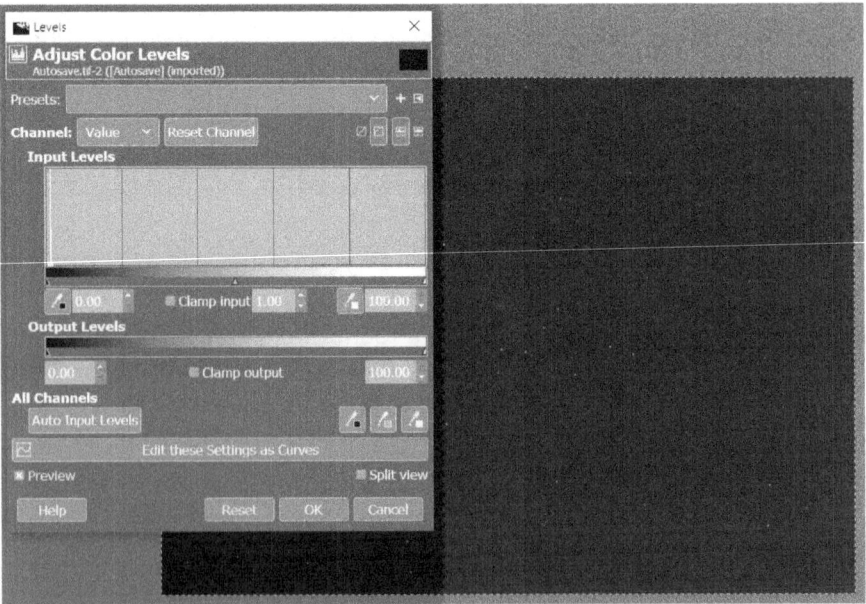

I like to zoom in to around 200% and find some bright stars. If you can find two stars close to each other, bring the slider over to a point before they blow out too much and become one starry blob.

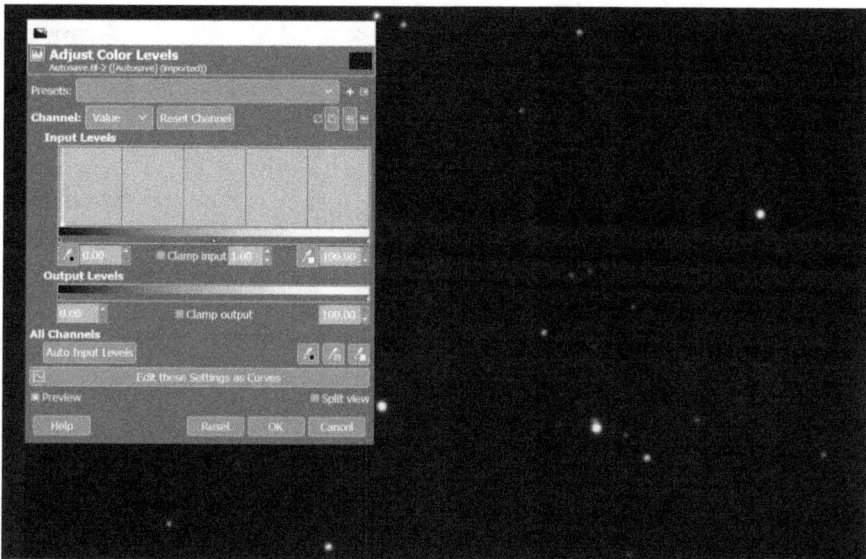

After the slider is adjusted…

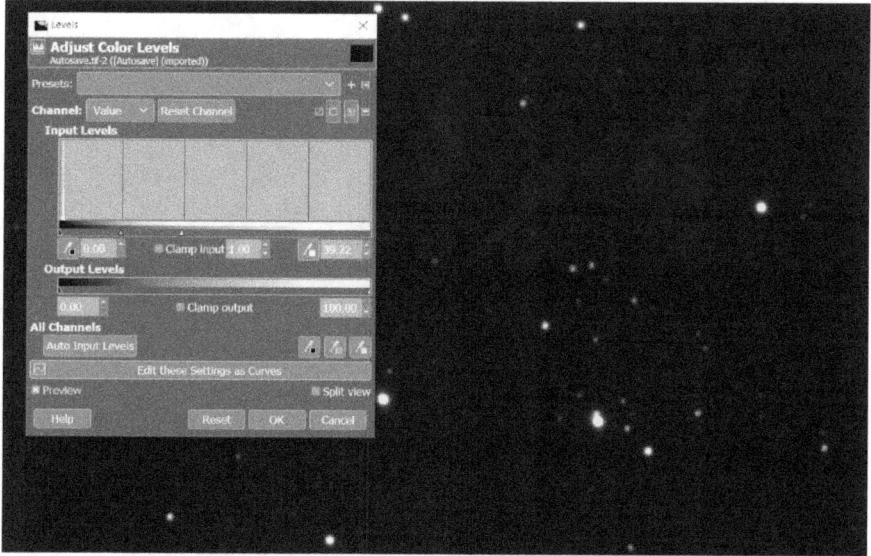

The next step I take is to adjust the middle slider and start to reveal the nebula. This shouldn't blow out stars but be careful. I would stop well before you hit the data mass.

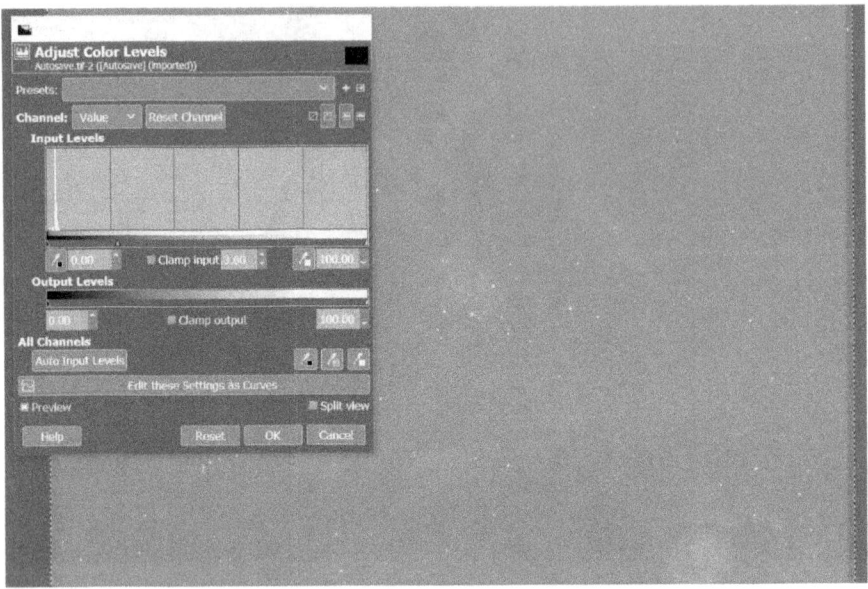

At this point, you can start to determine edge defects and determine where you want to crop. I crop at this stage.

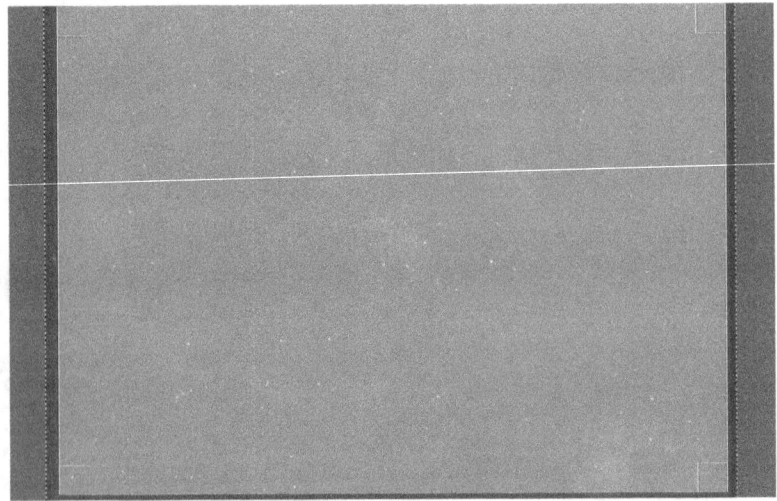

After cropping, I zoom in to ~200% again and adjust the left-most slider in levels. Here, I bring the slider all the way over to the data mass, but it reveals a good deal of noise after adjusting the middle slider again.

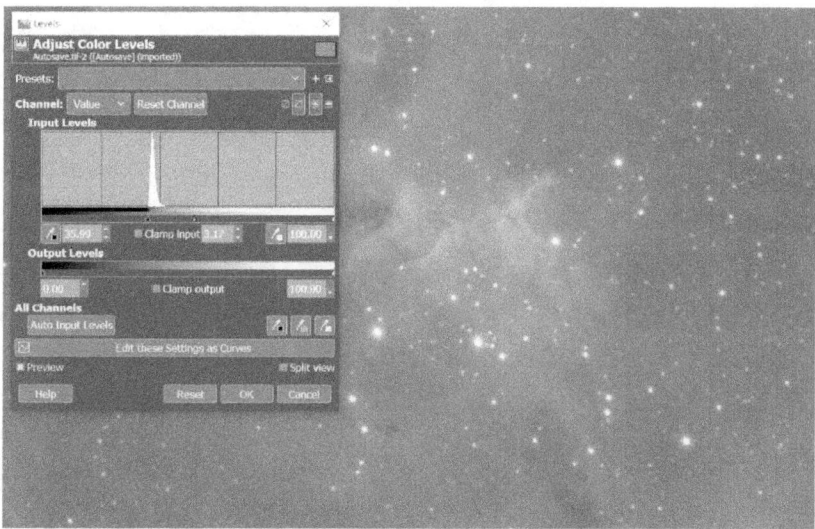

Here, I back off the left-most slider a little to lessen the noise.

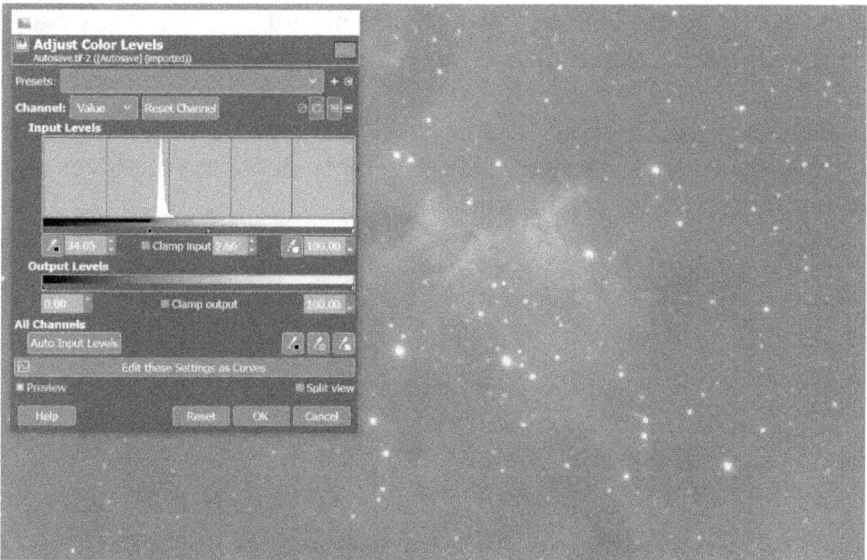

At this point, we can see the nebula and our stars are still tight and small. It is still quite dim at this stage.

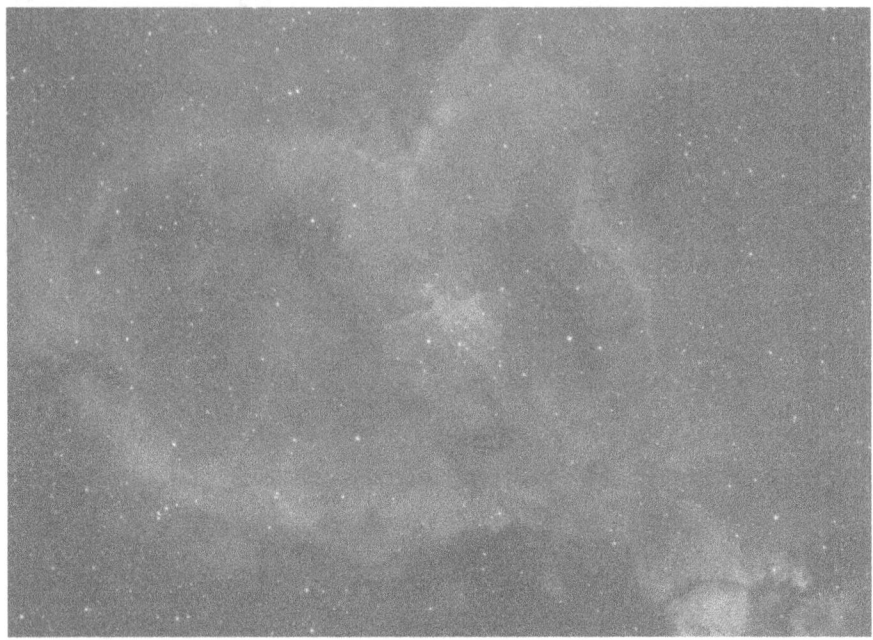

If you don't have any experience with photo-editing software, take a break and learn about layer masks. I will not dive into the details because I am no expert on the topic.

Now, we need to create a star mask to prevent our stars from blowing out. We want small stars! I simply duplicate the layer and re-name the two layers appropriately.

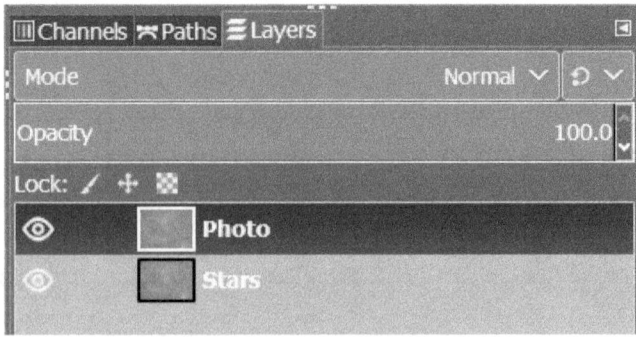

For this next step, I click on the "Stars" layer. Make sure it is selected.

Now, I zoom in to ~200% again and find a region full of bright stars. I then go to "select" > "by color" and select the centers of several of these bright stars. Adjust settings and selection points as necessary.

Now, I right click the "photo" layer and select "add layer mask" and chose "by selection" and check "invert mask". The following should appear under layers.

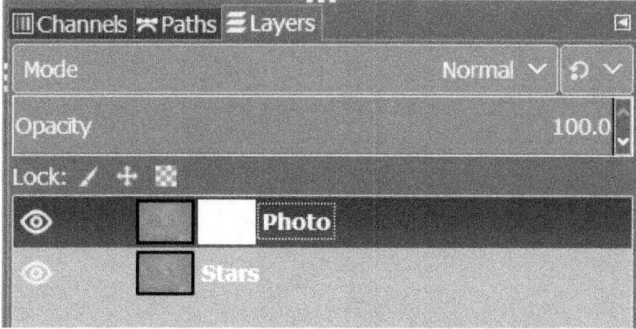

It is important that you see a white square pop up next to the photo layer, not a black one! Press the "ALT" key and click on the white square (the layer mask). The layer mask should pop up. You should see some faint black circles. These are your brightest stars.

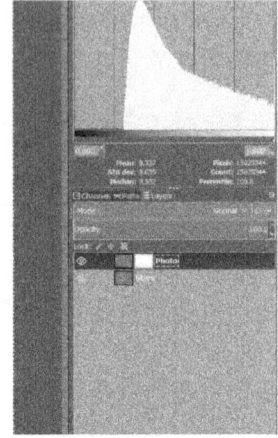

Now, adjust the left-most slider under levels and make these black stars more visible! Make sure the layer mask is selected, or else you will be editing something else.

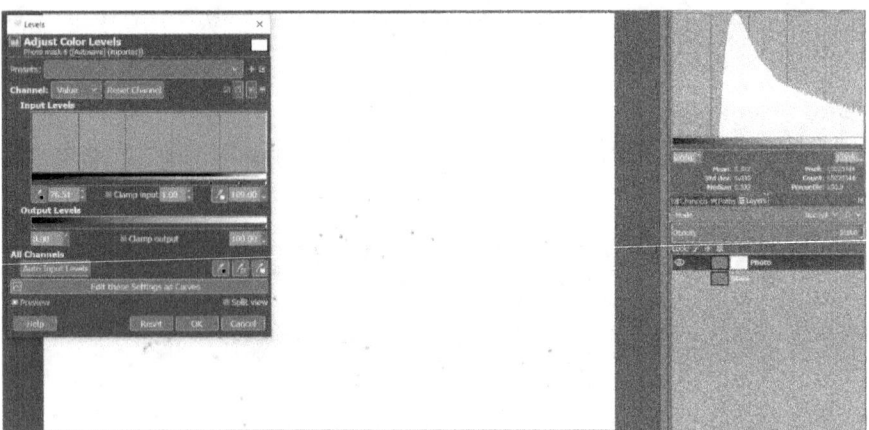

Now you can see the black circles!

Now it's time to blur the black circles a little. Blurring allows for the final image to be much smoother. You will understand why if you have a good understanding of layer masks. Before blurring:

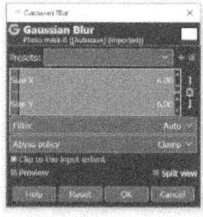
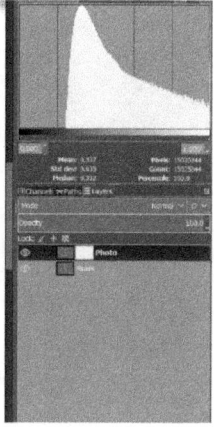

After blurring:

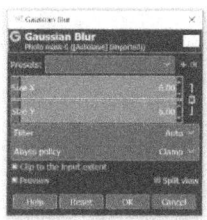
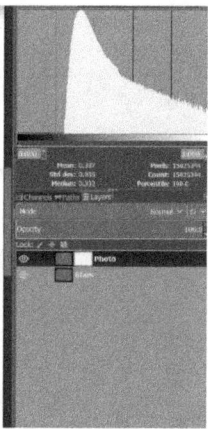

Nice, smooth transitions are preferable to harsh ones in photo editing. A soft mask will have soft looking stars, and this is a good sign that you have created a good layer mask for your stars.

Now, press "ALT" and click on the photo layer and get back to the main image.

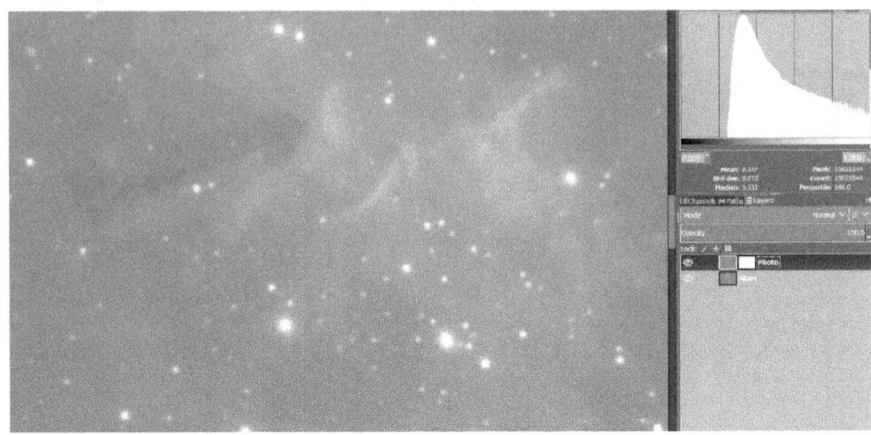

If you change the visibility of the "stars" layer, you can see the layer mask in action!

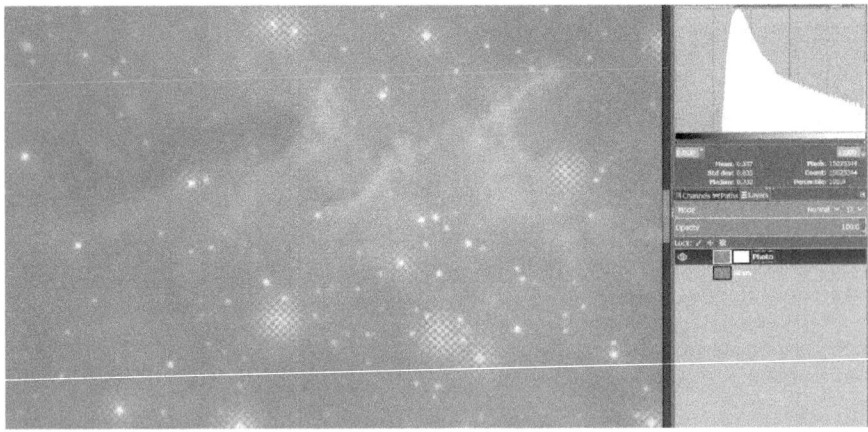

Here, you can see that all the brightest stars have become transparent and, therefore, will not be blown out any further when we brighten up the "photo" layer. Make sure opacity is at 100%. Make the "stars" layer visible again before continuing.

Now we can adjust "curves" and really bring out the detail in our image. Here is our status before the curves adjustment:

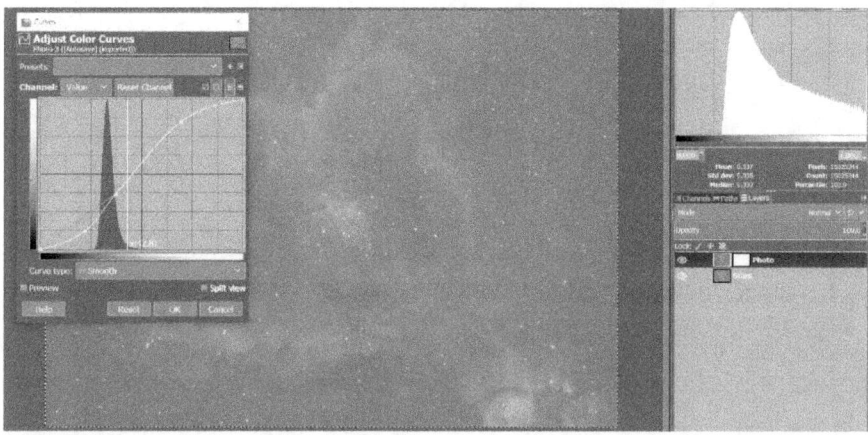

Here the result of the curves adjustment. I like to use a very organic-looking "S"-curve in curves, but the choice is yours.

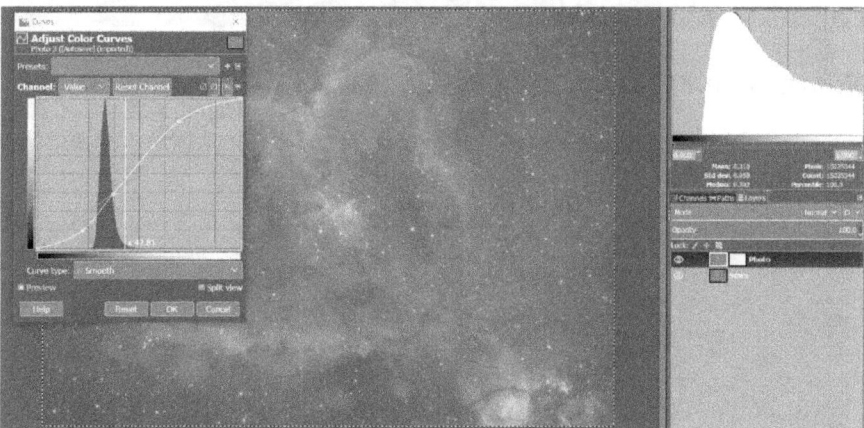

Unfortunately, I am unaware of how these images will come out in the print edition of this book, so I hope you are learning something!

Here is our status at this point.

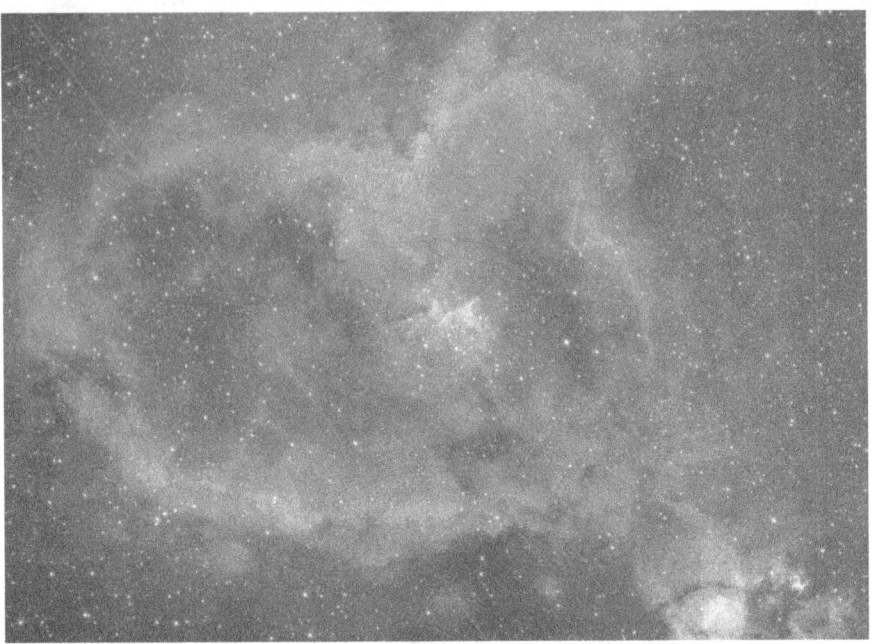

Now it is time to zoom in again and check on our stars.

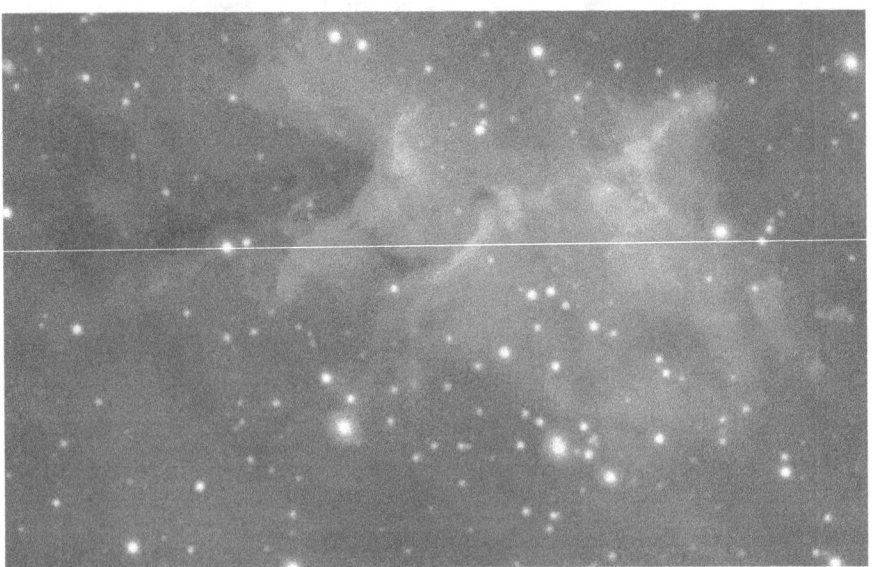

In this example, the stars still look quite nice. I will, however, act like there are some visible defects or harsh separations evident. A simple fix to odd star deformities or halos is usually to blur the star preservation layer mask some more. Again, press "ALT" and click the layer mask to change your selection. Make sure the layer mask is selected or else you will be blurring your final image!

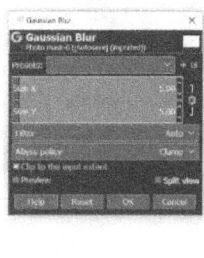
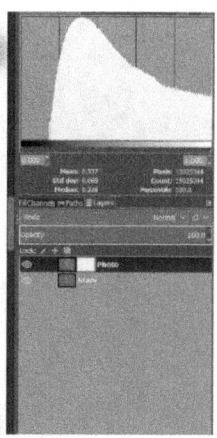

Here it is after some additional blurring.

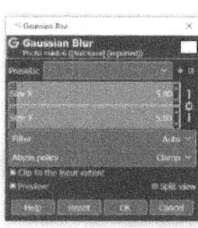
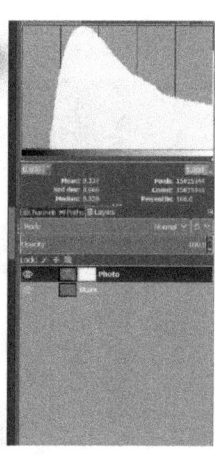

It should be obvious that the stars look a little softer.

Press "ALT" and click back over to the "photo" layer to get out of the layer mask. Zoom in and check up on the stars again. Hopefully any defects you saw earlier are gone. You can also make levels adjustments to the star preservation layer mask if additional blurring doesn't do the trick.

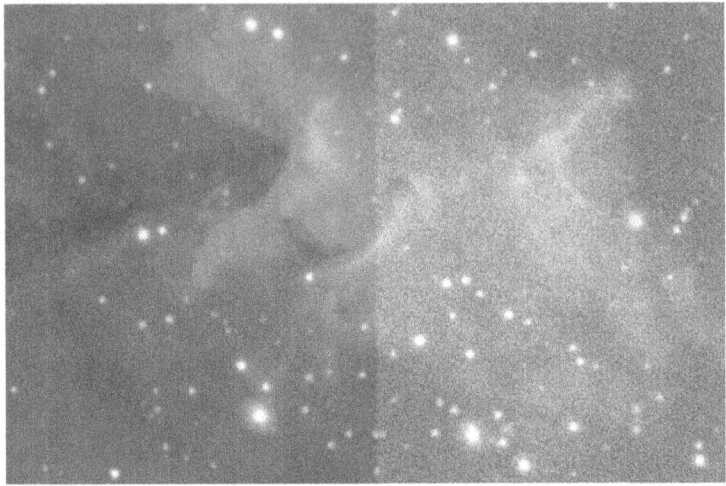

Here is the final result after some additional "shadows", "brightness", and "contrast" adjustments.

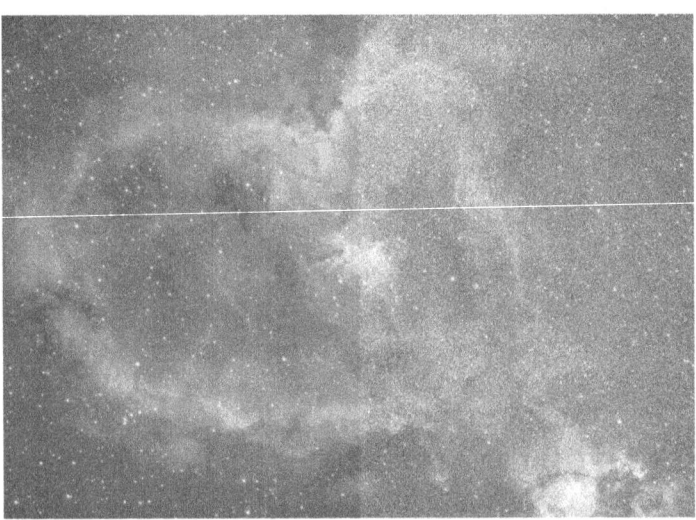

Chapter 13

Buying a Telescope & Camera

Most of the time, the amateur astronomer is "stuck with" a small refractor telescope (although I hope you realize by now that these are still impressive tools) they wanted to use to see if they were interested in pursuing this hobby. After one gets familiar with this "small" instrument and learns more about astronomy and telescopes; they are going to want a better instrument. It is human nature. When you go to buy your second telescope (or first telescope if you do not have one yet), you realize and ask yourself a few things.

(1) Do you want a refractor, a reflector, or something else? Reflectors typically have a much larger aperture and are, therefore, better for deep sky object observations. Refractor telescopes are generally smaller, but a 90mm is a very capable instrument. Reflectors are more complicated telescopes. Do you want a telescope that specializes in DSO observation or lunar and planetary observations? If you only want to observe DSOs visually, get a reflector. If you like all observations and want to stay cheap and simple, you should probably get a refractor with a larger aperture. The aperture of refractor telescopes is often reported in millimeters (70mm, 90mm, etc.), while the aperture of larger reflector telescopes is often reported in inches (6 inches, 8inches, etc.). The table on the next page lists some of the more popular aperture sizes and their millimeter or inch equivalent.

I have ignored binoculars in this book, but many people use them to observe the night sky. This book is about the small telescope, but the sky is the same for telescopes and binoculars. Binoculars certainly give you a wider field of view. The "# x #" in the description of astronomical binoculars is really "magnification by diameter of lens". Binoculars are for visual use only!

Popular Aperture (mm)	Popular Aperture (inches)
50	1.969
60	2.362
70	2.756
80	3.149
90	3.543
102	4.016
114	4.488
130	5.118
152.4	6
203.2	8

The "refractor vs. reflector" debate is quite a dangerous one to get involved in. Reflectors are usually larger, but refractor telescopes are more efficient. Refractor telescopes transmit most of the light (~90%) they collect into your eye. Reflector telescopes transmit less light (~70%), and this is due to the reflectivity of the aluminum coatings on the mirrors and the light blocked by the secondary mirror structures.

Reflectors are more expensive, but that is usually because they are larger and, therefore, more capable. Reflector telescopes, like most worth-while investments, require a decent amount of maintenance. They must be collimated (you will learn all about this if you buy one!) and the primary and secondary mirrors may have to be re-coated (very rarely). Collimation is not as intimating as it seems. If you want a reflector telescope, do not let collimation scare you away. You will be able to figure it out.

My rule? If you like to tinker, get a reflector. If you don't, get a refractor. Done. If you want your setup to always work, get a refractor. If you want to add a lot of integration time to your images over several nights, a refractor will be easier.

(2) What kind of mount do you want? Get an equatorial mount if you want to track objects. Whatever you choose, I strongly recommend getting a mount with those slow-motion control knobs. In fact, if I could, I would *require* you to get a mount with slow-motion control knobs. Remember that you can always weigh down your mount if it is not sturdy.

(3) How much do you want to spend? You can get very good telescopes for relatively low prices. If this is just a fun hobby to you, you do not need a $500 telescope. You can get very capable 70/80/90/102-millimeter refractor telescopes for under $200 (as of 2017). You can try used equipment as well. All your equipment is going to be "used" after a few nights anyway. Many people sell their 1.25" eyepieces after they upgrade to a telescope that requires 2" accessories. People generally take care of their equipment, so purchasing used equipment is not a terrible idea if you are on a tight budget and the return policy is good.

(4) What do I get next? One of the best things about this hobby is that it is not too hard on the wallet (if you stay an amateur and avoid astrophotography). Once you buy your second telescope, you may realize that you may not want to get anything else. All you need is a good telescope, some good eyepieces, and maybe a few good accessories. This is a unique trait of the "telescoping" hobby, so enjoy it.

Visual use only? If you like to tinker, get a large reflector (8 inches or larger). If you don't get a refractor (>90mm). A large fork-mounted SCT is another option for serious observers.

Astrophotography only? Start out with a 50-80mm f/5. Spend tons of money later (if necessary).

Light Gathering Power

Is there really a difference between a 50mm refractor and 90mm refractor? How about between an 80mm refractor and a 90mm refractor? The answer to the first question should be quite obvious – yes. There is a huge difference between a 50mm refractor and a 90mm refractor (you get what you pay for). A 90mm refractor will allow you to see DSOs that are seemingly invisible to a 50mm in light-polluted areas. A 90mm refractor is also much better for detailed planetary observations. Think about it. A 50mm refractor costs around $50 in 2017 while a 90mm refractor costs closer to $200. I own a 90mm refractor and a 50mm refractor and the 90mm refractor is far more enjoyable to use. That being said, a 50mm is still a perfect telescope for the beginner who likes a challenge. The answer to the second question is not so obvious. There is still a noticeable difference between an 80mm and a 90mm refractor. 80mm refractor telescopes are obviously incredible instruments, but 90mm telescopes still bring in more light. Instead of considering the diameter of the objective lens (the aperture), let's consider the area. The area of a circle is given by the following equation.

$$Area = \pi r^2 = \frac{(\pi d^2)}{4}$$

Table 14 (on the next page) lists the area of the objective lens of typical refractor telescopes using the equation above.

Telescope	Area of Objective Lens
50mm Refractor	19.63 cm^2
60mm Refractor	28.27 cm^2
70mm Refractor	38.48 cm^2
80mm Refractor	50.26 cm^2
90mm Refractor	63.61 cm^2
102mm Refractor	81.71 cm^2

Table 14 – The area of the objective lens of typical refractor telescopes. (100 mm^2 is equal to 1 cm^2)

A 90mm refractor's objective lens has a little more than 13 cm^2 of additional area than an 80mm refractor's objective lens. This additional area allows the 90mm to capture more light than the 80mm.

The light gathering power (LGP) of a telescope refers to how much more light the telescope can gather than a human eye. It is calculated as the ratio of the area of the objective lens to the area of the human eye. The diameter of a human eye can be estimated to be 7mm, but this diminishes with age.

$$LGP = \frac{(\frac{\pi d_{OBJ}^2}{4})}{(\frac{\pi d_{eye}^2}{4})} = \frac{d_{OBJ}^2}{d_{eye}^2} = (\frac{d_{OBJ}}{d_{eye}})^2 = \frac{d_{OBJ}^2}{49}$$

Table 15 lists the light gathering power of typical refractor telescopes using the equation on the previous page. For example, a 50mm telescope gathers 51 times more light than a human eye.

Telescope	LGP
50mm refractor	51.0x
60mm refractor	73.5x
70mm refractor	100x
80mm refractor	130.6x
90mm refractor	165.3x
102mm refractor	212.3x

Table 15 – Light gathering power.

In summary, do not be worried that you will not notice a difference if you buy a telescope with a slightly larger aperture. Table 15 shows us that the differences are quite dramatic. More light gathering power means you can observe fainter objects. This is important for DSO observation. Do not think that more aperture will eliminate light pollution, because it will not. If you live in a light-polluted area, light pollution is going to be a problem regardless of the telescope you use.

Field of view

I promised you a longer discussion on field of view so here we go. I did not find it appropriate to introduce field of view without discussing observations first. Whether or not you agree this discussion is appropriate here, we are discussing it here. The field of view dictates how much you see when you look through a telescope. Generally, the higher magnification you use, the less you will see. I believe that should be intuitive.

We have already discussed how some eyepiece are wide angle eyepieces because they let you see more. The *apparent field of view (AFOV)* refers to the field of view associated with the eyepiece. For example, say an eyepiece has an AFOV of 55 degrees. This tells you almost nothing about how much of an object you can see. The *true field of view (TFOV)* refers to how much of the sky you can see through an eyepiece.

$$TFOV = \frac{AFOV_{telescope}}{Magnification}$$

Since magnification has no units, the TFOV will be returned in degrees. For example, what is the TFOV I observe when I use my 90mm telescope with a 9mm eyepiece? (Magnification = 600/9 = 66.667) (Assume this eyepiece has an AFOV of 55 degrees)

$$TFOV = \frac{55}{66.667} = 0.82 degrees$$

Every object in the night sky has a size and this size can be expressed in arc-seconds or degrees. Table 16 lists the size of some DSOs. You can see why you cannot use high magnifications while observing deep sky objects. If you use too high of a magnification, the TFOV will be too small and you will not be able to see the entire object.

Object	Size (degrees)	Size (arc-minutes)
Full moon	0.5	30
M42	1.42 x 1.00	85 x 60
M31	2.97 x 1.05	178 x 63
M41	0.63	38
M44	1.59	95
M6	0.42	25
M7	1.33	80
M8	1.50 x 0.67	90 x 40
M20	0.47	28
M63	0.17 x 0.10	10 x 6
M27	0.13 x 0.09	8 x 5.7
M17	0.18	11
M1	0.10 x 0.07	6 x 4
M51	0.18 x 0.12	11 x 7
M13	0.28	16.6
M92	0.19	11.2
M65	0.13 x 0.03	8 x 1.5
M66	0.13 x 0.04	8 x 2.5
M57	0.02 x 0.016	1.4 x 1.0

Table 16 - Size of DSOs in arc-minutes and degrees.
(1 arc-minute is equal to 1/60 of a degree)

Tables 17, 18, 19, 20, 21, 22, 23, and 24 list the true field of view for several different telescope and eyepiece combinations. These tables should serve as more of a guide. They should help you estimate values if you do not feel like doing any math.

TFOV Tables

EP (mm)	AFOV				
	40	50	55	60	70
35	3.5	4.38	4.81	5.25	6.13
30	3	3.75	4.13	4.5	5.25
25	2.5	3.13	3.44	3.75	4.38
20	2	2.5	2.75	3	3.5
15	1.5	1.88	2.06	2.25	2.63
9	0.9	1.13	1.24	1.35	1.58
7	0.70	0.88	0.96	1.05	1.23
6	0.6	0.75	0.83	0.9	1.05
4	0.40	0.50	0.55	0.60	0.70

Table 17 – TFOV in degrees. f$_{OBJ}$: 400mm

EP (mm)	AFOV				
	40	50	55	60	70
35	2.8	3.5	3.85	4.2	4.9
30	2.4	3	3.3	3.6	4.2
25	2	2.5	2.75	3	3.5
20	1.6	2	2.2	2.4	2.8
15	1.2	1.5	1.65	1.8	2.1
9	0.72	0.9	0.99	1.08	1.26
7	0.56	0.7	0.77	0.84	0.98
6	0.48	0.6	0.66	0.72	0.84
4	0.32	0.4	0.44	0.48	0.56

Table 18 – TFOV in degrees. f$_{OBJ}$: 500mm

EP (mm)	AFOV				
	40	50	55	60	70
35	2.33	2.92	3.21	3.5	4.08
30	2	2.5	2.75	3	3.5
25	1.67	2.08	2.29	2.5	2.92
20	1.33	1.67	1.83	2	2.33
15	1	1.25	1.38	1.5	1.75
9	0.6	0.75	0.83	0.9	1.05
7	0.47	0.58	0.64	0.70	0.82
6	0.4	0.5	0.55	0.6	0.7
4	0.27	0.33	0.37	0.40	0.47

Table 19 – TFOV in degrees. f_{OBJ}: 600mm

EP (mm)	AFOV				
	40	50	55	60	70
35	2.00	2.50	2.75	3.00	3.50
30	1.71	2.14	2.36	2.57	3
25	1.43	1.79	1.96	2.14	2.5
20	1.14	1.43	1.57	1.71	2
15	0.86	1.07	1.18	1.29	1.5
9	0.51	0.64	0.71	0.77	0.9
7	0.40	0.50	0.55	0.60	0.70
6	0.34	0.43	0.47	0.51	0.6
4	0.23	0.29	0.31	0.34	0.40

Table 20 – TFOV in degrees. f_{OBJ}: 700mm

	AFOV				
EP (mm)	40	50	55	60	70
35	1.75	2.19	2.41	2.63	3.06
30	1.5	1.88	2.06	2.25	2.63
25	1.25	1.56	1.72	1.88	2.19
20	1	1.25	1.38	1.5	1.75
15	0.75	0.94	1.03	1.13	1.31
9	0.45	0.56	0.62	0.68	0.79
7	0.35	0.44	0.48	0.53	0.61
6	0.3	0.38	0.41	0.45	0.53
4	0.20	0.25	0.28	0.30	0.35

Table 21 – TFOV in degrees. f_{OBJ}: 800mm

	AFOV				
EP (mm)	40	50	55	60	70
35	1.56	1.94	2.14	2.33	2.72
30	1.33	1.67	1.83	2	2.33
25	1.11	1.39	1.53	1.67	1.94
20	0.89	1.11	1.22	1.33	1.56
15	0.67	0.83	0.92	1	1.17
9	0.4	0.50	0.55	0.6	0.7
7	0.31	0.39	0.43	0.47	0.54
6	0.27	0.33	0.37	0.4	0.47
4	0.18	0.22	0.24	0.27	0.31

Table 22 – TFOV in degrees. f_{OBJ}: 900mm

EP (mm)	AFOV				
	40	50	55	60	70
35	1.40	1.75	1.93	2.10	2.45
30	1.20	1.50	1.65	1.80	2.10
25	1.00	1.25	1.38	1.50	1.75
20	0.80	1.00	1.10	1.20	1.40
15	0.60	0.75	0.83	0.90	1.05
9	0.36	0.45	0.50	0.54	0.63
7	0.28	0.35	0.39	0.42	0.49
6	0.24	0.30	0.33	0.36	0.42
4	0.16	0.20	0.22	0.24	0.28

Table 23 – TFOV in degrees. f_{OBJ}: 1000mm

EP (mm)	AFOV				
	40	50	55	60	70
35	1.27	1.59	1.75	1.91	2.23
30	1.09	1.36	1.50	1.64	1.91
25	0.91	1.14	1.25	1.36	1.59
20	0.73	0.91	1.00	1.09	1.27
15	0.55	0.68	0.75	0.82	0.95
9	0.33	0.41	0.45	0.49	0.57
7	0.25	0.32	0.35	0.38	0.45
6	0.22	0.27	0.30	0.33	0.38
4	0.15	0.18	0.20	0.22	0.25

Table 24 – TFOV in degrees. f_{OBJ}: 1100mm

I really struggle to recommend DSLRs for astrophotography anymore. If you have one at home, use it. If you don't you really should buy at least a cheap dedicated astronomy camera. If you have a DSLR without live view, I wouldn't bother using it.

In DSO astrophotography, common guidance is to aim for a resolution between 1-2 arc-seconds per pixel. Getting that resolution is difficult for beginners though, so we will leave it at that. For wide-field imaging, which is recommended for beginners anyway, that 1-2 arc-second rule really doesn't have relevance.

If you want to worry about this or have a little more experience, you should buy a camera with a pixel size that works with your telescope and yields ~1-2"/pixel resolution.

Planetary & lunar imaging can become quite a fun conversation. There are several ways to optimize your system for peak planetary/lunar performance. **First, I will present what the common advice usually is**:

1. For normal nights, aim for 0.25"/pixel. Aim for an optical system F-number of 5x your camera's pixel size.
2. For nights with excellent seeing, aim for 0.1"/pixel. Aim for an optical system F-number of 7x your camera's pixel size.

It's good advice, but I like to think about it a little more. This all has to do with *sampling* which is a topic I am adamant about avoiding in this book. I will not talk about it. **Here is the thought process I like to follow:**

Let's think about your telescope. We already know how to calculate the resolving power of a telescope.

$$R\ (arcseconds) = \frac{120}{Aperature\ (mm)}$$

For example, a 10" telescope has a resolving ability of 0.47". I think we can all agree. Some heavy math tells us that it is best to cut this in half that way the smallest detail able to be resolved by the telescope is covered by two pixels. I think this makes logical sense if you think about it. That means we should aim for ~0.24"/pixel.

Now, let's solve the following equation:

$$Focal\ Length\ (mm)\ =\ \frac{pixel\ size\ (um)}{R_I\ ("/pixel)} * 206.3$$

Let's consider a camera with a 2.9um pixel size and a 10" telescope.

$$2{,}492mm\ =\ \frac{2.9um}{0.24"/pixel} * 206.3$$

In this case, we should aim for ~2500mm, which is a F/10 system. If you have a 10" F/5 Newtonian, put on a 2x barlow. If you have a 10" F/10 SCT, don't worry about a barlow. Instead, crop the video size to maximize your camera's frame rate. There are many ways to over-think this, but I like this method the best. Here's a table with common apertures for this style of imaging:

Aperture (mm)	Resolving Ability (")	Sampling Goal ("/pixel)	Pixel Size (um)	Target Focal Length (mm)	Target F Number
			2.9	1516	10
152	0.79	0.39	3.5	1829	12
			5	2613	17
			2.9	2024	10
203	0.59	0.30	3.5	2443	12
			5	3490	17
			2.9	2533	10
254	0.47	0.24	3.5	3057	12
			5	4367	17
			2.9	3041	10
305	0.39	0.20	3.5	3670	12
			5	5243	17

How to buy a dedicated astronomy camera for DSO use

1. Color or Mono?
    a. Only you can answer this question.
2. Cooling
    a. Cooling is needed.
3. Diagonal
    a. You need to determine the image circle of your telescope or field flattener / coma corrector. Thanks to wonderful marketing strategies, you may want to use a sensor with a sensor diagonal smaller than the listed image circle by a good margin.
4. Pixel Size
    a. You should aim for an image scale of 1-2"/pixel. Anything smaller than 1"/pixel is going to require significant effort, but it is still very possible. It depends on the style of astrophotography you are interested in. An image scale of 1.5"/pixel is still interesting.

$$R \ ("/pixel) = \frac{pixel\ size\ (um)}{Focal\ Length\ (mm)} * 206.3$$

5. Megapixels (MP)
    a. MPs are slightly over-rated but anything over 20MP will give very high-quality results.
6. Quantum Efficiency (QE)
    a. This is becoming irrelevant now with camera sensor technology getting so good. Still peak QE of 80% is incredible.
7. Dynamic Range
    a. 12 stops is outdated, 14 stops is a good goal.
8. Full Well Capacity
    a. 25ke is good, but 50ke is better. Full well capacity relates to how quickly you clip data by over-exposing bright regions.

DSO Astrophotography Field of View

Both photos were taken using the same camera. The first was taken using a telescope with 250mm focal length. The second was taken using a telescope with 442mm focal length. The focal length preference is yours. Of course, you can even use a longer FL!

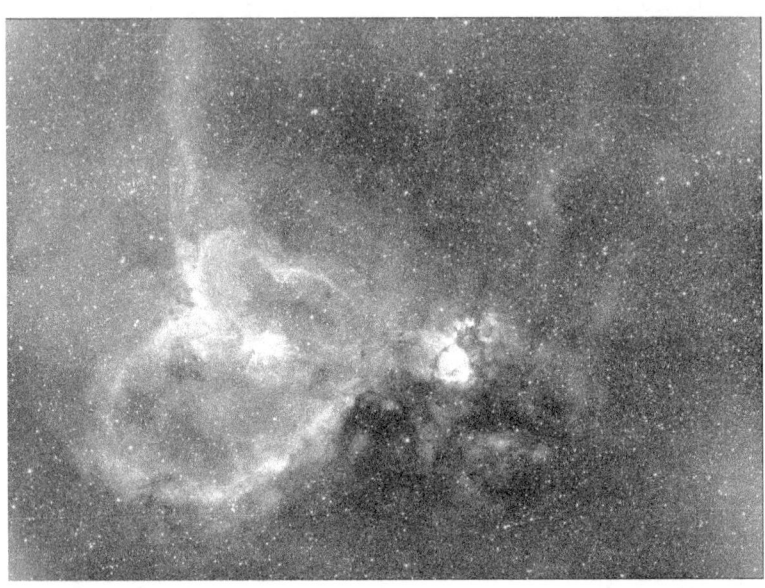

SCT Economics

You probably wouldn't be able to guess from reading this book, but I love using SCT-type telescopes for planetary & lunar imaging. They also make wonderful visual scopes when mounted on a fork mount.

Let's consider the popular sizes of 6" (150mm), 8" (200mm), 10" (250mm), and 12" (300mm). Let's look at the prices and compare the target resolution for planetary and lunar imaging. These prices are a rough estimation around the year 2020.

Aperture (in.)	Aperture (mm)	Resolution (")	Target Res. ("/pixel)	Price ($USD)
6	150	0.80	0.40	1000
8	200	0.60	0.30	1800
10	250	0.48	0.24	2400
12	300	0.40	0.20	2900

Let's plot the data.

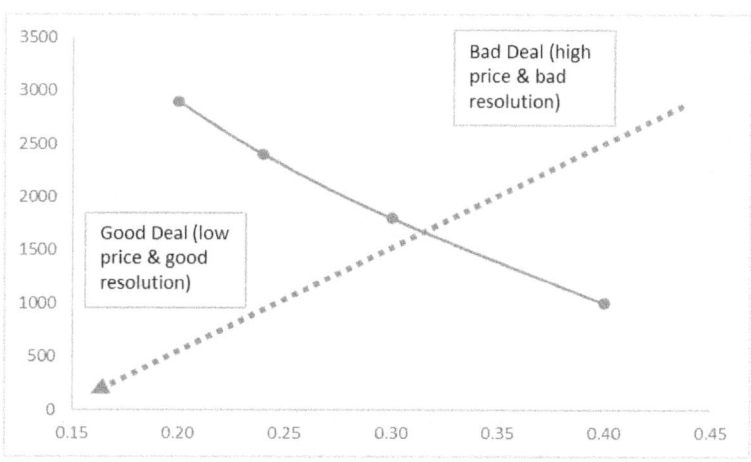

I'll admit, I was a little shocked to see how linear this data was. It seems like there is no obvious best choice. Maybe a little economy of scale, but nothing obvious.

This data was generated using the 6" mounted on a single arm fork mount and the other larger scopes mounted on a double arm fork mount. This is standard in the industry.

However, since we are only interested in visual and planetary and lunar imaging, we don't need fantastic stability. We can, therefore, find 8" SCTs mounted on single arm fork mounts that would be good for our purposes. We do not have this option with the larger scopes. We can re-do our analysis with the 8" on a single arm fork.

Aperture (in.)	Aperture (mm)	Resolution (")	Target Res. ("/pixel)	Price ($USD)
6	150	0.80	0.40	1000
8	200	0.60	0.30	1200
10	250	0.48	0.24	2400
12	300	0.40	0.20	2900

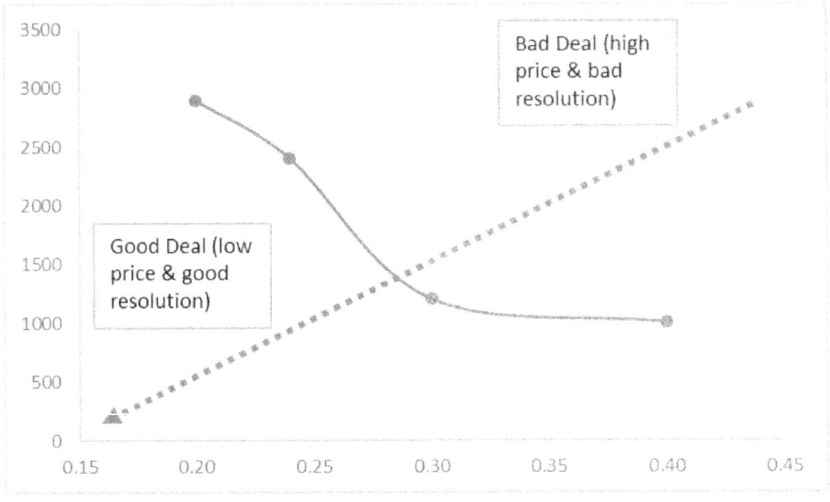

Now the 8" SCT stands out a little more and looks like a good balance of price and resolution. This is just a very simple example of some fun telescope economics.

Part 4: Additional Topics

Here, we are going to cover a few "bonus" topics you may or may not be interested in.

Chapter 14

Equatorial Mount Basics

Declination and right ascension are the coordinate system astronomers use to define the location of celestial objects in the sky. They are relatively confusing and intimating. They are also not that helpful to many people (in fact, I rarely use them).

Your latitude on Earth is the same as the declination of a star or object that passes directly over you. For example, on the northern pole, the latitude is 90 degrees and Polaris is directly overhead; therefore, the declination for Polaris should be close to 90 degrees, which it is (89° 15' 51"). Declination describes the position of an object relative to the celestial equator.

Right ascension is slightly more confusing. Basically, astronomers have decided on a reference point and an object's right ascension is reported in hours away from that reference point. Although this is not too important, understanding the general idea behind the celestial coordinate system is a good skill to have. Equatorial mounts keep a telescope pointing at a constant declination. Right ascension (RA) motors, keep a telescope pointing at a constant right ascension. Combined, these two will keep an object still in the view of the telescope since both the declination and right ascension are locked in.

These values are current as of 2000, but they change over time.

Star	Declination	Right Ascension
Polaris	+89° 15' 51"	2h 31m 49s
Sirius	-16° 42' 58"	6h 45m 9s
Vega	+38° 47' 1"	18h 36m 56s

German amateur equatorial mounts are very intimating at first, but they can make your astronomy life much easier. First, you should familiarize yourself with the basic parts of a German equatorial mount.

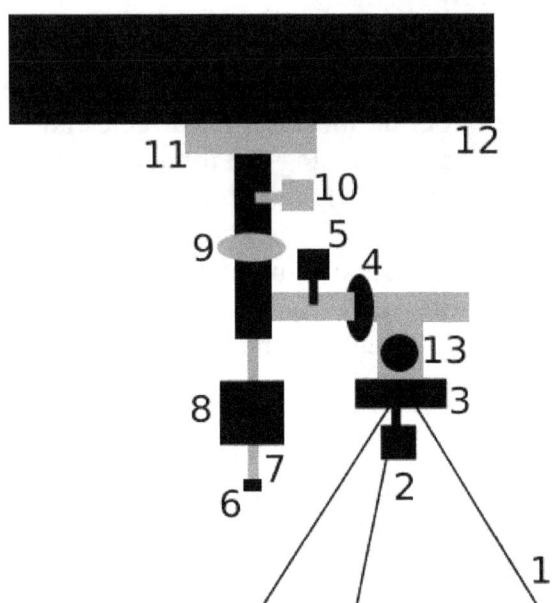

1	Tripod	8	Counterweight
2	Azimuth Lock	9	Declination Setting Circle
3	Mount Base	10	Declination Lock (Clutch)
4	Polar Setting Circle	11	Telescope Mounting
5	Polar Lock (Clutch)	12	Telescope
6	Counterweight Stop	13	Latitude Adjustment
7	Counterweight Shaft	13	Fine Latitude Adjustments

You should also familiarize yourself with the two axis of your equatorial mount. Practice moving your telescope about each axis. It may help to lock one of the axis, so that you can tell when you are really only moving about one axis.

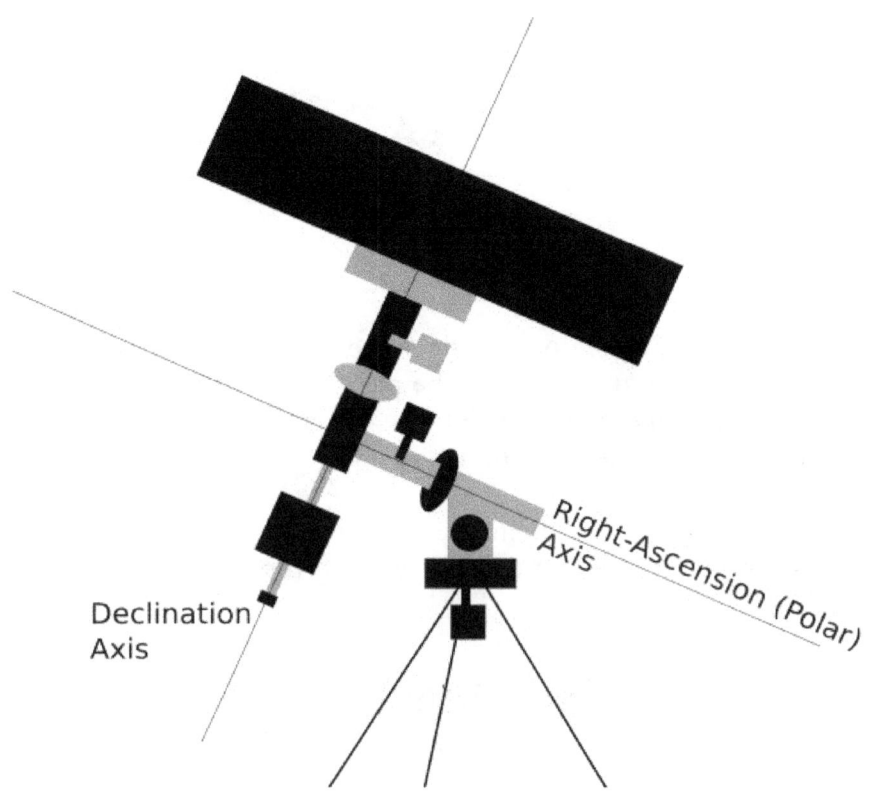

Right-Ascension Axis (Polar)

Declination Axis

Before you can start using your telescope, you should practice balancing the entire instrument. You may have to balance over and over again throughout the course of the night. If you add a smartphone to the telescope, you have to balance again; if you add a Barlow lens, you have to balance again; even if you change eyepieces, you may have to balance again. Do not fear, you can balance your telescope in two simple steps. <u>You can probably do a rough balance and be okay, but it is good to exaggerate at first.</u>

First, make sure everything is at 90 degrees, just like the first image in this chapter. Loosen the declination axis. Rotate around the polar axis until the counterweight shaft is parallel to the ground. Lock the Polar axis. Adjust the position of the telescope in the rings so that the telescope does not dip forward or backward.

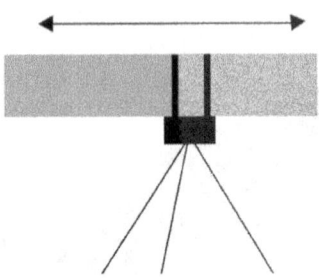

Loosen the polar axis. Lock the declination axis at 90 degrees. The counterweight shaft should still be parallel to the ground. Adjust the position of the counterweight on the axis until the telescope does not drift in either direction.

Please remember to tighten the counterweight throughout this process so that it does not fall off the shaft (do not trust the safety nut).

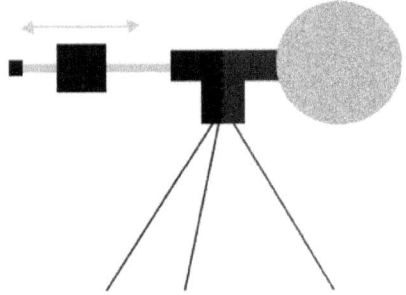

After you complete these steps, loosen both the declination and polar axis. The telescope should not drift in any direction no matter what declination and right-ascension point you use. Lock the axis when you are done.

Many equatorial mounts can overcome slight load imbalances, but you should still try to get the best balance you can. Equatorial mounts tend to perform much better when they are under loaded. For example, if your mount is rated for 12lbs. and you load it with 5lbs. of accessories, it will perform much better than if you loaded it with 11lbs. of accessories.

In order for your equatorial mount to function properly, you have to align it with your pole. If you live in the northern hemisphere, you have to align in with the northern pole.

If your mount does NOT have a polar axis finder, you have to trust your telescope. Mount your telescope with both the declination and right ascension axis locked at 90 degrees. Adjust the latitude setting until it matches your current latitude and swivel your mount head until Polaris is centered in your eyepiece or camera. When you do this, your mount is roughly aligned.

If you mount does have a polar axis finder, simply look through the finder and align with the northern pole. The polar axis finder will come with more detailed instructions, but they are very easy to use.

In order to get the best possible alignment, you need to perform a drift alignment or some other similar alignment method. I usually stick to the simple, one-star alignment using my mount's polar axis finder.

Once you align your mount, you have to lock the latitude adjustment and the mount's head nut. The only movement should be done through the unlocked right ascension and declination axis. Only adjust the axis when they are unlocked. Do not make harsh adjustments with the axis locked. You use fine adjustment knobs when the axis are locked. The RA motor should only be used when the RA axis is locked.

An equatorial mount with RA motor will enable you to take long exposure photographs of DSOs or longer videos of planets and the moon. A dual axis controller makes things easy because you can use a hand controller to adjust the RA and declination axis.

In the astrophotography world, sometimes people purposely use a slightly unbalanced setup. The reason for this is relatively simple to understand: gears are not perfect! Slack between teeth on gears can cause shaking and ruin an image. To overcome this, you can create a slight load imbalance to "help" the RA motor.

For example, consider the following two scenarios. In both examples, your mount has already been aligned and you are ready to shoot. Imagine your telescope is sitting between you and Polaris (or whatever pole you are aligning with).

(1) If you point to an object to the left of Polaris, your countershaft is likely going to be the left and your telescope is likely going to be to the right. In your position, the RA motor is going to turn your RA axis counter clockwise. In other words, the countershaft is going to drop and the telescope is going to rise. In this case, astrophotographers can decide to make the counterweight shaft side slightly heavier (by shifting the counterweight down just a little). Now, the extra weight is helping the motor and putting less stress on the motor. This will create smoother tracking.

(2) If you point to an object to the right of Polaris, your countershaft is likely going to be the right and your telescope is likely going to be to the left. In your position, the RA motor is still going to turn your RA axis counter clockwise. In other words, the countershaft is going to rise and the telescope is going to drop. In this case, astrophotographers can decide to make the telescope-side slightly heavier (by shifting the counterweight up just a little). Now, the extra weight is helping the motor and putting less stress on the motor. This will create smoother tracking.

In less words, astrophotographers tend to make whatever side is towards the rotation of the RA axis slightly heavier to possibly eliminate the shakes associated with gear slack. This becomes fairly intuitive the more you get acclimated with your mount.

Computerized Go-To Mounts

A wonderful technological revolution in astronomy has led to the development of computerized Go-To mounts and even computerized polar alignment.

Every high-end equatorial mount I can think of has go-to capability. It makes astrophotography much easier. In order to use this functionality, you must perform some star-alignment procedures that can be found in your owner's manual. You don't even need to do these star alignments anymore if you have plate solving ability. The technology is getting ridiculous!

What is more important for this section of the book is that polar alignment is getting easier too. You can purchase polar alignment cameras that make accurate polar alignment much faster.

A rough alignment may be all that is necessary for visual astronomy or planetary and lunar imaging. If a rough alignment is quicker, simply complete a rough alignment for those activities.

My Current Equatorial Mount

I now use a center-balanced equatorial mount (CEM) instead of the traditional German equatorial mount (GEM). The CEM design is generally much lighter than a GEM for a given mount capacity. Also shown here is the portable tripod-pier hybrid that I recommend for extra stability.

Chapter 15

Autoguiding

Astrophotography is a pretty bad addiction.

Many people believe that...

You are not really imaging until you start guiding.

And, in a way, they are right. We need very long exposures, right? Some narrowband imagers shoot 1-hour sub exposures. An amateur unguided equatorial mount cannot handle that. Periodic errors (among others) start to affect the image.

The solution? Guiding. What is guiding? Guiding refers to the process of "locking" a certain star in a position in the field of view. Guiding can fix periodic errors and allow for very long exposures.

Manual guiding refers to the process of looking through a very long focal length telescope and adjusting the RA and declination fine adjustment knobs to keep a star fixed in a location. No one does this anymore for good reasons.

Everyone autoguides. Autoguiding puts this task on a computer and allows us to relax and enjoy the night. You can probably guess that this can get extraordinarily expensive and you would be correct.

So how do we do this in a budget-friendly way? Use a webcam and a tiny telescope as a guidescope. This can get ugly fast, but the general process is not too complicated.

Typical setups are shown in the following images.

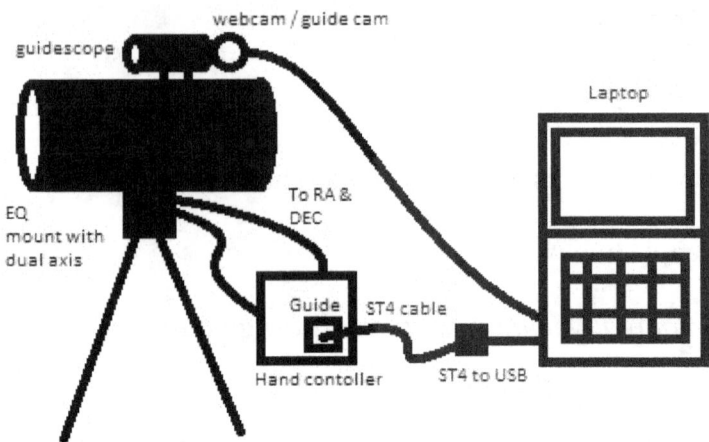

Method 1: Use a cheap webcam. We need a bright field of view to get a bright enough guide star. Use a fast guide scope. The webcam plugs into the laptop (easy). Connect a ST4 cable to the guide port on your mount. Use some type of ST4 to USB conversion to plug into your laptop. Use free autoguiding software (PHD2 is a good program at this time). Choose a bright enough star and done!

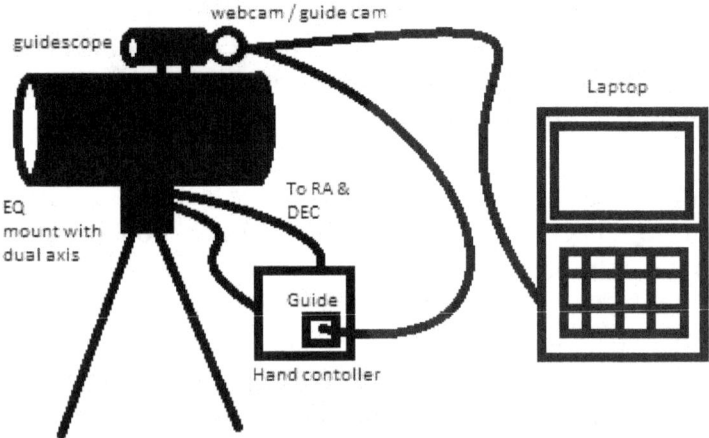

Method 2: Use a dedicated auto-guiding camera. Similar to method 1 besides connections.

Your guide camera must be compatible with your computer. Your unique setup is going to require unique drivers and your own research, but it does not have to be incredibly complex.

Choosing a guide scope and guide camera

It is my opinion that there is a lot of bad advice out there about choosing a guide scope. It is simple. The guide scope and guide camera work together – they cannot be separated.

Advice I hate: "Your guide scope focal length should be at least 1/10 focal length of the imaging scope"

Advice I hate a little less: "Your guide scope image scale should be no more than 3 times greater than your main imager image scale"

Image scale is very important here. It is all that matters. The focal length of your scope and the pixel size of your camera determine your image scale. Divide your pixel size by your focal length and multiply by 206 to get the image scale. You need to consider both your guide scope AND your guide camera.

My advice: Guide at a scale that matches your seeing if you are imaging at focal lengths greater than 400mm. Done.

If you are imaging at 300mm, guide at whatever image scale you want to. A little 30mm guide scope should be plenty.

Autoguiding programs can guide at resolutions of 1/10 pixel. I don't like to rely on this too heavily. If your seeing conditions typically allow resolutions in the 2" range (OK seeing) then your guide scale should be 2"/pixel. I think 2"/pixel should be a minimum for everyone.

Consider a star. It is a point source of light. They are very small on a camera's sensor. If seeing permits 2"/pixel and you are guiding at a scale of 6"/pixel (I have done it), you will get some movement within a single pixel that the computer would not be able to recognize. That movement would destroy a nice image if the image scale is 1"/pixel. Even worse, the computer would think the guiding accuracy is great because it can't determine any errors.

It is easy to imagine how a guide scale of 6"/pixel would produce blocky stars. That is not what we want. If we get closer to a guide scale of 2"/pixel, the star will be spread out over more pixels, thus making it easier (or even possible) for the computer to guide at 1/10 pixel accuracy. At 2"/pixel, stars are rounder.

Advice for imagers is to image at image scales between 1-2"/pixel. I say guide at the same resolution for the same reason – stars are round (well-sampled) at 2"/pixel. We want round stars in our images, but also in our guiding. It may be difficult to get 2"/pixel with common guide scope and guide camera combinations.

Guide Scope Focal Length (mm)	Guide Camera Pixel Size (um)					
	2.5	3	3.5	4	5	6
50	10.3	12.4	14.4	16.5	20.6	24.7
100	5.2	6.2	7.2	8.2	10.3	12.4
150	3.4	4.1	4.8	5.5	6.9	8.2
200	2.6	3.1	3.6	4.1	5.2	6.2
250	2.1	2.5	2.9	3.3	4.1	4.9
300	1.7	2.1	2.4	2.7	3.4	4.1
400	1.3	1.5	1.8	2.1	2.6	3.1
	Guiding Image Scale ("/pixel)					

To make it even worse, it really isn't ideal to mount a 300mm guide scope on your imaging telescope since it'll just make the whole setup that much heavier. In my opinion the best way to guide is off-axis guiding. (It also makes balancing a Newtonian much easier if you plan to image with a Newtonian)

Off-axis guiding allows you to guide at the same focal length of your main imaging scope without the weight of an additional guide scope. This involves a prism that picks a little light away from the light what would have missed your camera's sensor anyway and directs it towards a guide camera. Setting up an off-axis guider takes a little bit of work, but the benefits greatly outweigh the struggles of the initial setup.

Here is a diagram showing how this works on a refractor telescope.

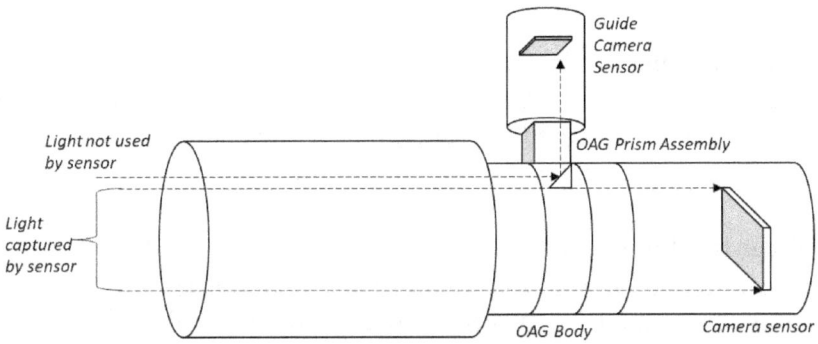

Here is another diagram that more clearly shows how the light that the guide camera uses would otherwise not be used by the imaging camera. A great way to ensure there is no interference is to take flat frames and see if the prism casts a shadow on those. The prism should not cast a shadow if it is set up properly.

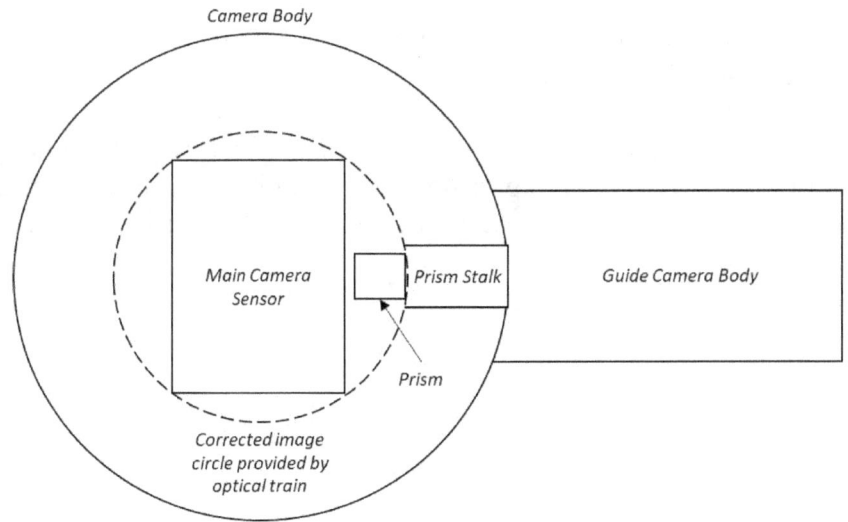

In a very nerdy way, off-axis guiders may be a great way to extract the maximum value out of your optical train. Your telescope (and other associated optics) already provides a corrected image circle. You *paid* for that entire image circle, despite you only using a rectangular sensor with a diagonal close to the diameter of the corrected image circle. An off-axis guider allows you to use some of that "wasted" corrected image circle. How cool is that?

Differential flexure

It has been argued many places that differential flexure is a far bigger problem than guider image scale. Differential flexure occurs when you use a separate guide scope and the guide scope can move independently from the main imaging scope (even if very slightly). It becomes a much bigger problem with longer focal lengths.

Differential flexure may be the bigger problem. I could be completely wrong in my approach to recommend a guider image scale of 2"/pixel minimum for autoguiding, but the advice is still the same – get an off-axis guider.

An off-axis guider eliminates differential flexure, eliminates the problem of guider image scale, eliminates the need for a separate guide scope, eliminates the need for a dew strip for the separate guide scope, and so much more. It is a great way to guide, regardless of your imaging scale. Yes, the initial set up is a little annoying, but once you've done it, you should not need to adjust it.

Off-axis guiding setup

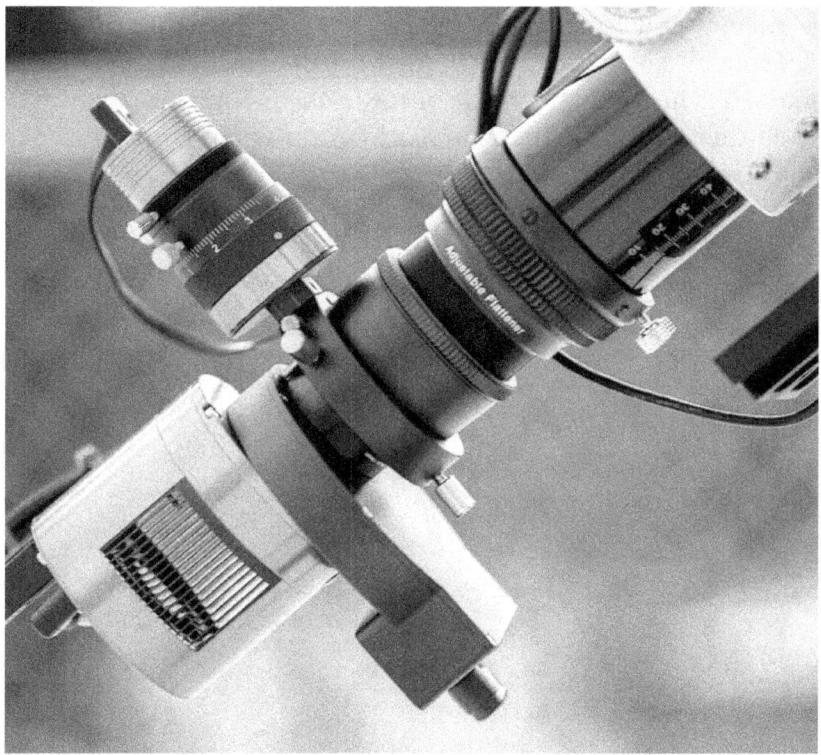

Here, you can see the main monochrome imaging camera connected to a filter wheel. A 10mm M42 spacer separates the filter wheel and the off-axis guider body.

The guide camera is focused by rotating the helical focuser. Off-axis guiding is something that appears very complicated until you commit yourself to figuring it out. Once you do, it is a very nice way to accomplish good guiding!

Off-axis guider setup

The raising and lowering of the internal prism depends on your exact model, so that cannot be discussed. However, what can be discussed is how much to raise or lower the prism. Take flat frames and do this during the day. Be sure not to completely loosen the prism so it can't fall loose.

Flat frame #1 (prism too low):

After raising the prism slightly, you get flat frame #2 (still too low):

After raising the prism slightly again, you notice that the shadow disappears. You have correctly setup your off-axis guider.

Off-axis guiding is particularly useful with imaging reflectors. Balancing these telescopes is almost impossible if you auto guide with a separate, smaller guide scope. Off-axis guiding is very useful in this case.

Separate guide scope makes balancing difficult:

Off-axis guiding is a clean solution for reflectors especially.

Chapter 16

DIY Telescopes & Accessories

Why would anyone want to make their own telescope?

There are many people who make their own giant telescopes. Honestly, if you ever get to the point where you feel confident making your own 22 inch reflector telescope, you are not going to be referencing this book!

First, let us discuss tiny telescopes. Why would you want to make your own tiny telescope? Two reasons.

1. You want to build your own guidescope for autoguiding.
2. You want to build a unique telescope with specific traits.

Making smaller telescopes is relatively easy. Since you are building a smaller telescope, it should probably be a refractor. To make your own refractor telescope, all you need is an objective lens (with cell), a focuser, and a tube. I built my own 70mm f/4 telescope and I will show you how. I use it as a guidescope but also sometimes *attempt* to use it visually. With a focal length of only 280mm, it gives some fun views. I call it, "miniscope". It is obviously extremely portable and looks more like a camera lens.

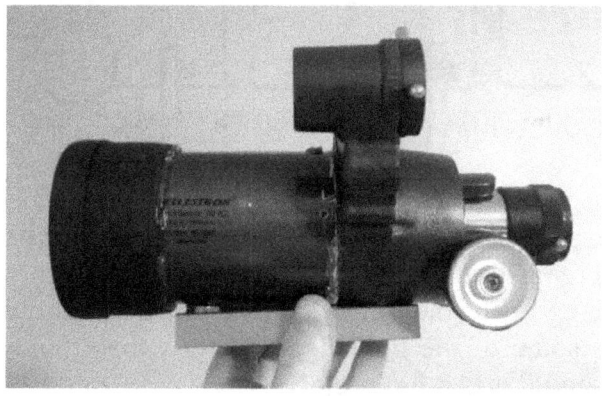

How did I build it?

1. I purchased a 70mm f/13 telescope at a flea market in horrible condition. What did I care? I only needed the tube, the focuser, and the dovetail!
2. I purchased a used 70mm f/4 objective lens (in a cell) from an old pair of binoculars.

I cut the tube until a rough assembly was able to focus on a star. Once achieving focus was possible, I reassembled the parts. The following two images show this assembly.

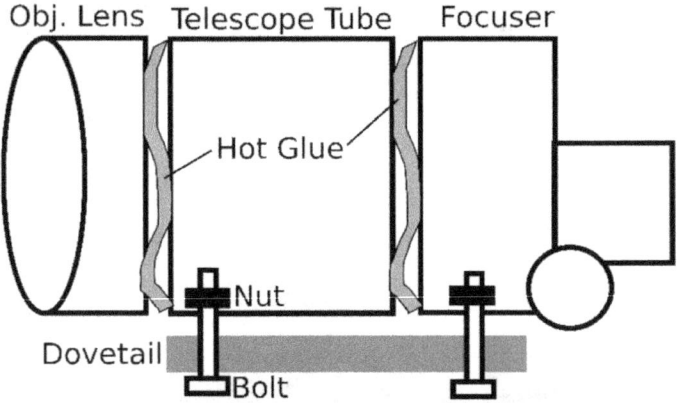

Perhaps the biggest issue with this design is the fact that the focuser is designed for a 70mm f/13 not a 70mm f/4. I had to cut the focuser tube on the end near the objective lens but I am sure the focuser still blocks some of the light. Despite that, the "miniscope" still works fine and it was a fun project.

You may think you are unable to build your own larger Dobsonian type telescope. Trust me, you can. Building your own Dobsonian telescope is a lot of (easy) fun. It may or may not save you money depending on what aperture you choose to build and what supplies you already have at home.

If you look this topic up, you will find a lot of great telescope builds. Instead of adding to this large list of "great Dobsonian telescope builds", I am going to show you how to build a very cheap Dobsonian telescope. Sure, it may not be pretty and it may not be as easy to use, but it works.

First, I am going to over-simplify this "price" topic just to make a point. Building a Dobsonian with an aperture greater than 10 inches will likely save you a little money. Building a Dobsonian with an aperture under 10 inches will likely cost more. This is not always true, but it is a good rule to keep in mind. Either way, it is very rewarding and fun to build your own. Don't do it to save money.

Keep in mind that making your own optics is almost always going to cost you more money these days. Buy your primary and secondary mirror already made.

Design choices

(1) What aperture? I did a 10" f/5 project once using a very crude construction. I do not recommend the following construction, but it is a nice thinking exercise for you if you have never thought about making your own telescope before.

(2) Dobsonian or dovetail? Dobsonian is fun to build but buying a dovetail may enable you to mount your telescope on an existing mount and save you money. Weight may be an issue on an existing mount. If you are going greater than 6 inches, I would go Dobsonian unless you have some experience.

(3) Truss or solid tube? Solid bodies act as a giant dew cap for your mirrors. And block stray light. Truss-types look awesome and the mirrors may cool to ambient temperature faster. I chose truss type. I don't care what anyone says. I love a well-designed truss type telescope.

As I mentioned previously, if you are seriously building your own larger Dobsonian telescope, you will probably not be referencing this book. I am simply going to provide you with some notes on the exact construction path I went down.

Notes on construction:

(1) I used a jigsaw. Many people demand you to use a router but if all you have is a jigsaw, it can be done. Jigsaws are cheap if you do not have one.

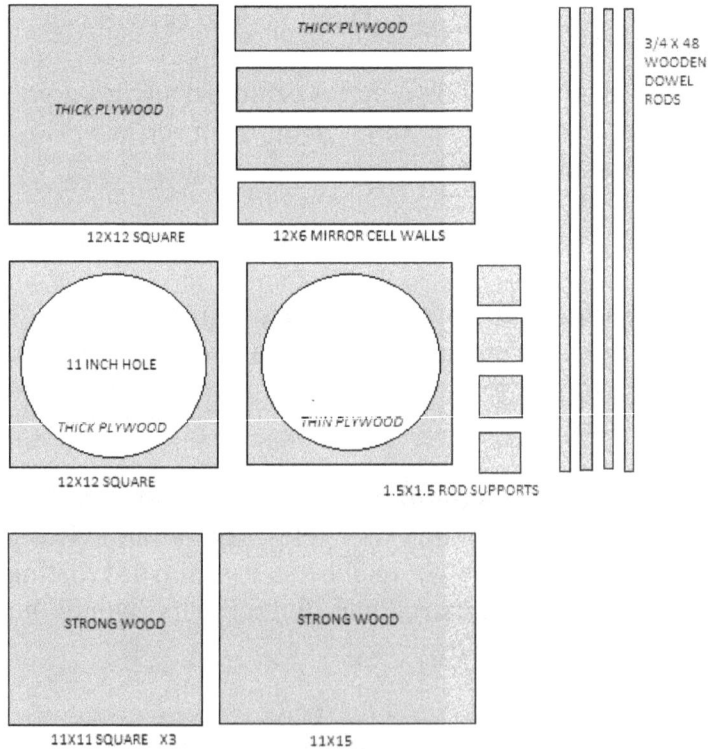

(2) I coated all my wood with polyacrylic to help it resist the outdoor elements a little better. Polyurethane is also an option.

(3) I purchased a 10inch circle from an arts and crafts store to mount the primary mirror on. I cut an 8inch hole in the center. You can make a primary mirror cell as follows. This will allow you to collimate the mirrors. There should be a gap between the mirror and the cell for air to pass through. Silicone will fix the mirror to the cell, not the mirror clip. You can also buy mirror cells if you want to skip this step.

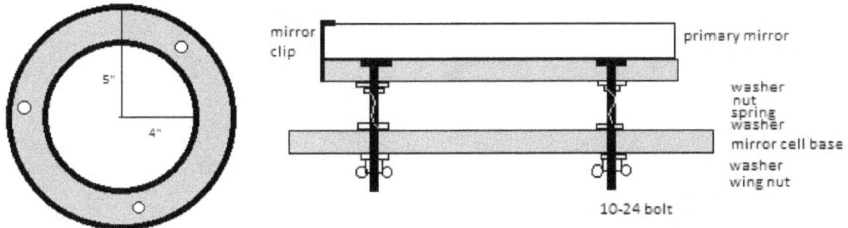

(4) I purchased a cheap "lazy Susan" from a discount store.

(5) I purchased 2 ABS (PVC would work too) 4 inch drain plugs.

(6) For the secondary mirror holder I purchased one 24 inch 10-24 bolt and a single 1" dowel rod. I used the thicker plywood "12x12 square with 11 inch hole". I used an epoxy adhesive to attach the secondary mirror to the rod.

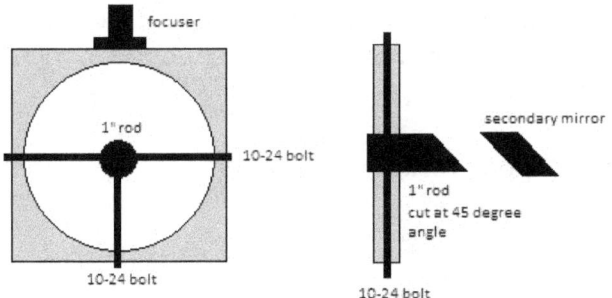

(7) I used a combination of wood glue, screws, and brackets. You want a strong base and a strong overall construction!

Final construction as follows…

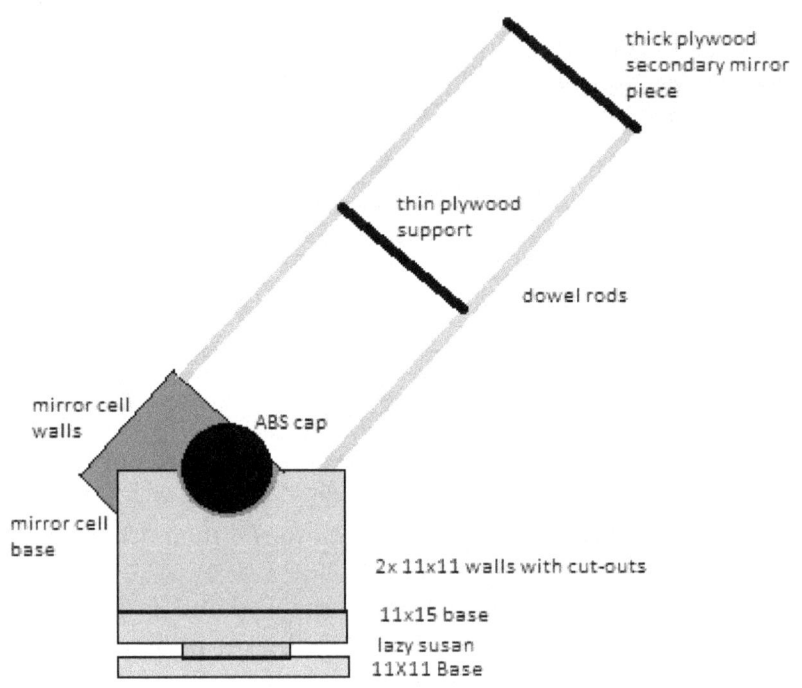

You will likely adapt/improve this design to better suit your needs. This is, however, the very basic structure of a DIY Dobsonian telescope.

Fans are a very good idea!

I also completed a 6" f/8 project. I started to hate "wood-working" after my 10" f/5 project. Metal is better suited for me. This telescope blew me away the first time I used it! I loved the thing. I built this one on an all-aluminum angle bar and flat bar truss structure. Very strong and over-engineered, but nice.

I set out to build a simple, lightweight, strong, and affordable Newtonian telescope. Of course, I think everyone wants a telescope with all these features. Naturally, I couldn't get all 4. I just wanted to build an effective reflector that most people would be able to find all the parts for and build easily.

I really wanted a long-focus planetary killer. With affordability in mind, I went with a 6" f/8 primary and 31mm secondary. This should be a 20% obstruction and, therefore, should be good visually. With simplicity in mind, I bought the primary box from an arts and crafts store. I also limited the construction to wood and aluminum bar. With lightweight in mind, I went with an open truss, aluminum bar set up. Total weight is just under 12lbs. With Strength in mind, I went with the truss/aluminum angle construction. I mounted this scope on a leftover AZ mount.

Here is a close-up view of the very basic secondary holder. I limited myself to compression springs, #6 bolts, and 1" wooden square dowel. Two of the three bolts have springs and can be adjusted. The third bolt is there just to keep it sturdy. I find this works quite well actually. The fan cools the secondary and keeps dew away. This is just another design idea for you. Your creativity is the limit.

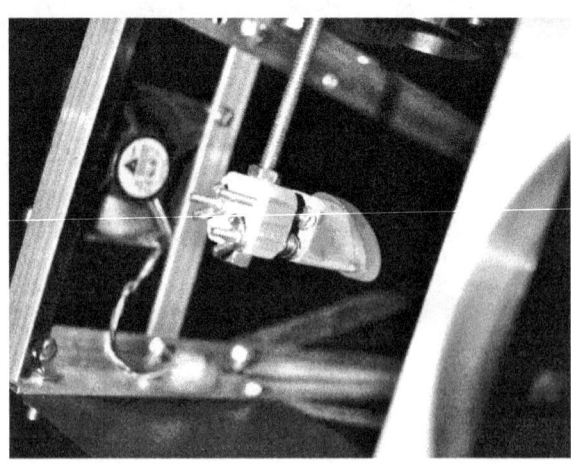

I also designed a nice 6" f/6 design on a computer aided design program. I haven't had the time to build it yet, but maybe you do!

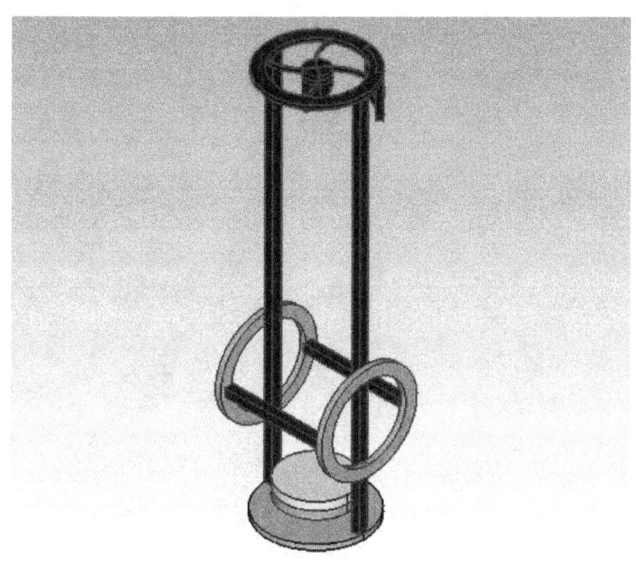

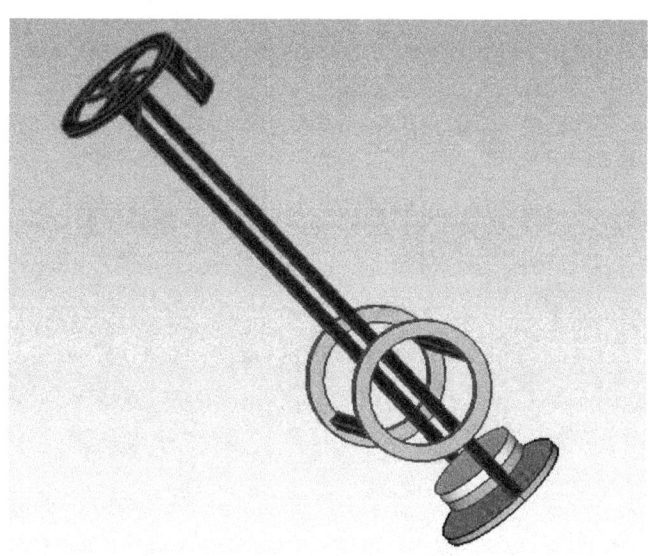

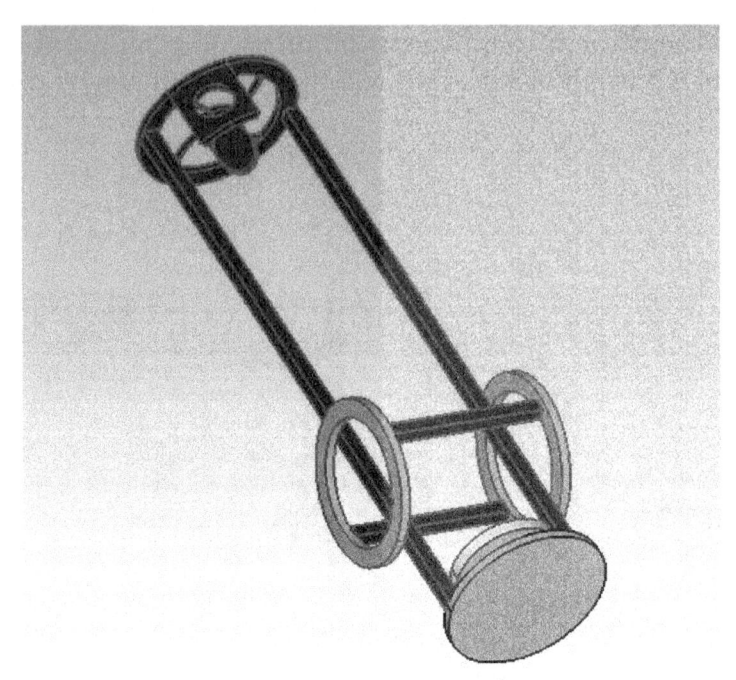

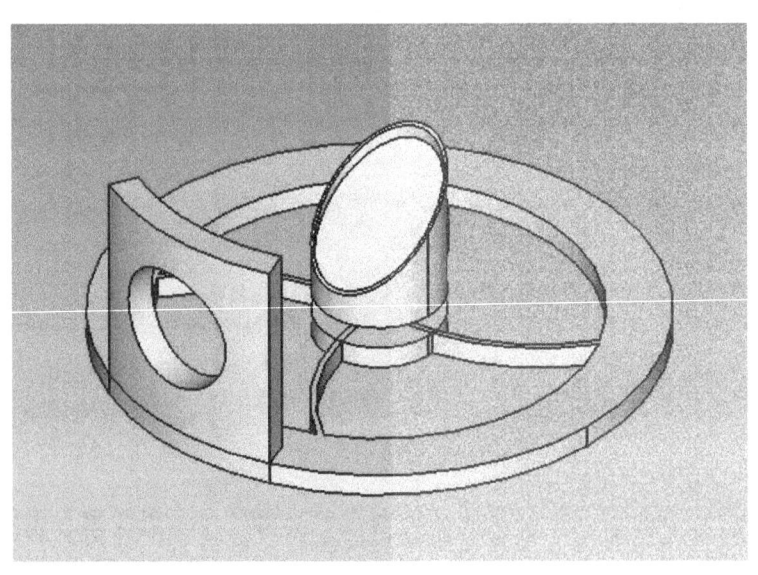

DIY Field Battery

A field battery can power a telescope mount, a camera cooler, dew strips, you name it. A DIY 12V field battery is a great way to get portable power for your astronomy setups.

The Battery

12V deep cycle marine battery is best. The amount of amp-hours (Ahr) you want depends on your equipment. Unless you have a ridiculous amount of equipment, 36Ahr is a good starting point. What this means is that you can draw 1 amp for 36 hours, 2 amps for 18 hours, 3 amps for 12 hours, etc...

Your battery is going to have positive and negative terminals. It is very important to note that raw batteries do not usually have electronics built it. There is no regulation, it is just simple chemistry. Be careful.

We need to put a fuse somewhere after the positive terminal. Common advice is within 10 inches of the terminal. You want the fuse close to the terminal to reduce the risk of a short. If a short occurs, nothing will stop the flow of electricity besides the wire burning up. This is very serious. For most normal setups a 7.5amp fuse is good.

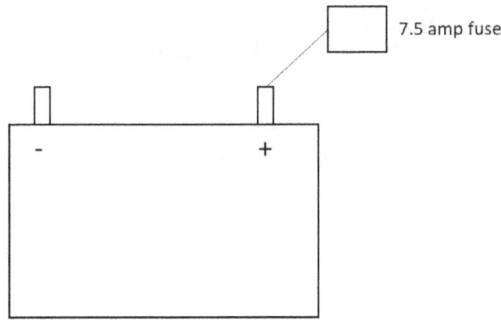

7.5 amp fuse

Now you need to choose your electrical connector of choice. I prefer the 2.1x5.5mm connectors because they are very cheap and very common (splitters/adapters are also readily available). Then we just have to connect everything as follows.

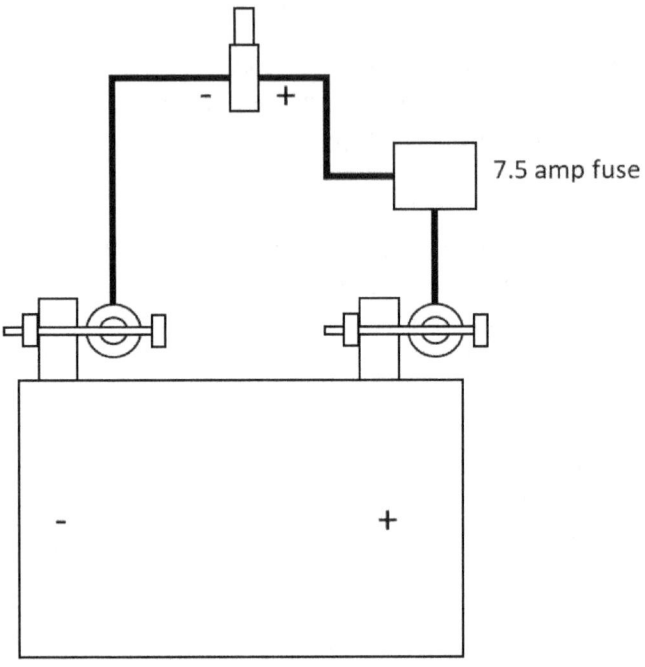

It is good practice to put heat shrink over any crimp style wire connectors you use. I always feel safer that way.

If you mount requires a 5.5mm male plug, you just have to order the proper extension cable to connect your mount to your battery. If you want to power multiple devices, get a splitter cable.

You will need a battery charger to recharge your field battery after you deplete it. It may take more time than you would like to re-charge it to full capacity.

Chapter 17

Reflectors and Collimation Principles

Only the Newtonian reflector will be considered as it is rare amateur astronomers purchase the other variations. A Newtonian reflector telescope is essentially a tube with two mirrors: the primary and the secondary. While refractor telescopes are usually ready to use immediately after opening the box, reflector telescopes may require some mirror alignment. The process of aligning the mirrors properly is called collimation. A collimated telescope will perform much better than a telescope with a poor alignment. In other words, you cannot simply ignore collimation.

Collimation is confusing and scares some people away from buying reflector telescopes, but it is not all that horrible. You can buy tools that help the collimation process, such as a Chesire eyepiece or a laser tool. In order to use a Chesire eyepiece, the center of the primary mirror must be marked in some way. You can also collimate with no tools at all.

Since we have established that you are likely on a budget, we will collimate without any tools. Please note that collimation will most likely be better using the tools mentioned above, but we can do well without them. A star test is the best anyway.

Usually the mirrors are already aligned when you receive your telescope. My 130mm reflector telescope came with a good collimation, but I messed it all up so I could show you how to do it. I do not recommend messing with the mirror alignment if you received the telescope with proper collimation.

Step 1 is to remove the cap on the focuser tube and look into it with your eye only. You will see a few things.

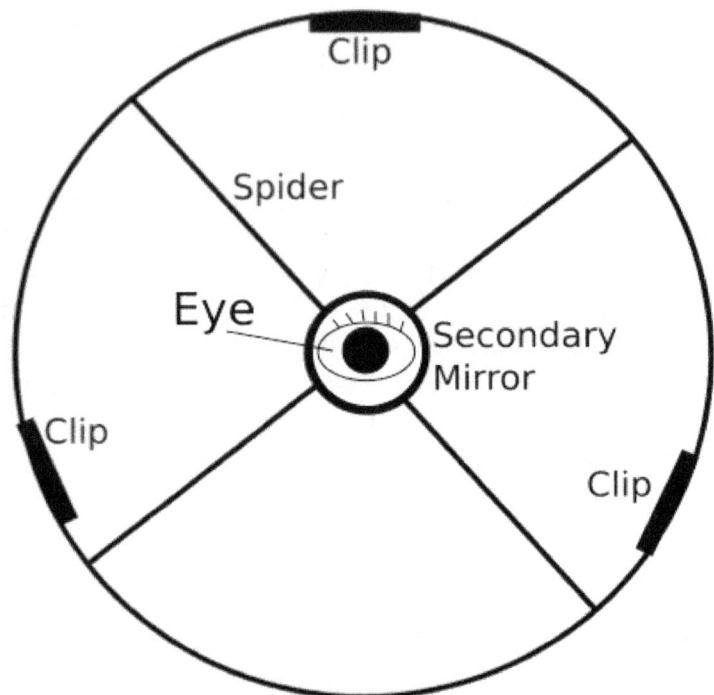

Figure 16.1 – Looking into the focuser tube of a reflector.

You are going to want to do this in a room with the lights on. Point the telescope at a bright wall in the room. In order for this method to work, your eye has to be centered in the reflection on the secondary mirror. If you see something exactly like figure 16.1, then you are most likely good to go. You are still going to want to confirm the collimation with a star test.

Figure 16.2 – Misaligned secondary mirror.

In figure 16.2, the secondary mirror is horribly misaligned. You will see a few (probably three) secondary mirror adjustment screws on the back of the secondary mirror holder. Adjust these screws until you see the entire primary mirror and the three clips that hold it in place. Do not worry about anything else for now. Figure 16.3 shows an approximately correctly aligned secondary mirror. Yes, it really is that easy.

Figure 16.3 – Correct secondary mirror alignment.

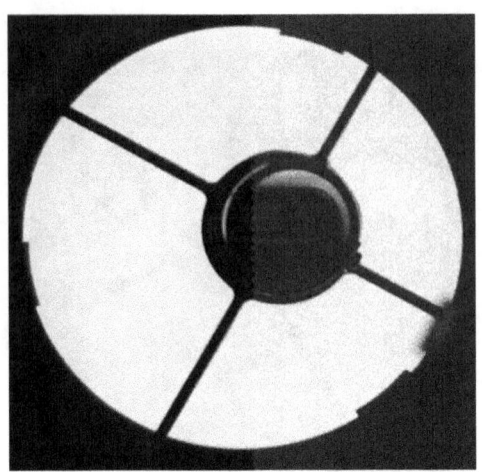

Figure 16.4 – Misaligned primary mirror.

In the image above, the primary mirror is misaligned. Your goal is to approximately center the secondary mirror in the primary mirror. There are screws at the mirror-end of your reflector telescope. Adjust these screws (read the instruction manual to find out which ones they are) until you have accomplished the image in figure 16.5.

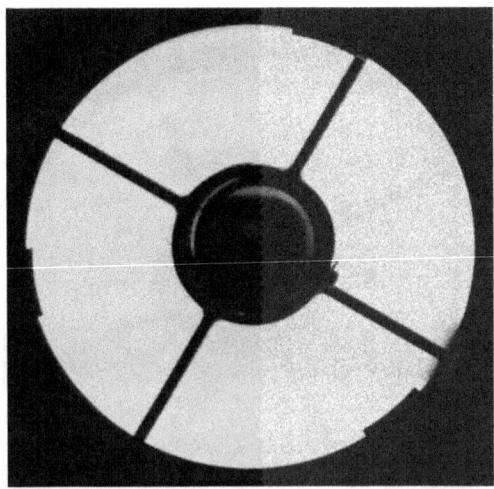

Figure 16.5 – Correct primary mirror alignment.

At this point, your reflector telescope is now approximately collimated. Again, I must say that this collimation is only approximate and that your eye must always be centered in the secondary mirror. You really should purchase some tools once you get more serious about this hobby. I like to use an expensive laser.

To confirm the collimation, you must do a star test. A star test is quite simple. Point your telescope at Polaris. Polaris is an ideal choice because it is roughly stationary, but any second or third magnitude star will work just fine. You should use lower magnifications for this.

Focus the star in the center of the field of view. Now slightly take the star out of focus (in any direction – it does not matter). You should see bright rings of white light form around a black circle (the secondary mirror obstruction).

If the black circle is centered in these bright white rings, you are good to go.

If the black circle is not centered in these bright white rings, you have some more alignment to do. First, move the out-of-focus star around the field of view until the black circle becomes centered in the rings of white light. Now, adjust the primary mirror collimation screws until the out-of-focus star is centered in the field of view. Once you have done this, you can re-focus the star. You may wish to take this star out of focus again and repeat the procedure until it appears to be properly aligned. It is not unusual to have to do this more than once.

If you want to collimate the proper way, get yourself a high-quality & expensive laser collimator or a standard collimation cap. This rough method will at least get you started though. I currently use a Cheshire eyepiece and finish with a star test.

Chapter 18

Budget Astronomy

Budget astronomy using a small telescope can actually be quite fun. You do not need to drop thousands of dollars on a very large reflector telescope to get great views of the night sky. Your budget is your budget. If you can only comfortably afford a 70mm refractor telescope, get it. There is no reason for such a fun hobby to cause financial stress in your life.

Budget astrophotography is impossible, but very fun. You cannot take incredible pictures of deep sky objects without accurate tracking and long-exposure photography using a digital SLR camera or dedicated astro-camera. That being said, look at the images in this book. Most images were taken using "cheap" telescopes. They are nothing special, but they are still photographs you can show off.

If you have the money and this is something you really enjoy, please go ahead and spend the large amount of money required for professional astrophotography. If this is just a fun hobby for you (like it is for me), do not be afraid of the cheapest setup you can possibly build and spend some time working on it and tweaking it yourself.

So what do you really need?

Visual:
1. You need a telescope. Whatever you can afford. You should really consider a larger Dobsonian. Consider a DIY Dobsonian.
2. You need a mount. Altazimuth or Dobsonian if you want to stay cheap and simple.
3. You need quality eyepieces.

Lunar/ Planetary Astrophotography:
1. Telescope.
    a. Large Newtonian provides big aperture (great detail) for low cost.
    b. Small refractor provides high contrast views and low maintenance but with chromatic aberrations. Get f/10 or slower (f/11, f/12, etc...).
    c. Cassegrain provides long focal length in a small package but can be more complex to maintain.
2. Mount.
    a. Dobsonian for larger Newtonians. Motorization enables longer videos.
    b. Equatorial with tracking for long refractor or Cassegrain allows for longer videos.
3. Camera
    a. Webcam.
    b. DSLR with crop mode.
    c. Dedicated astronomy camera

DSO Astrophotography:
1. Small, fast refractor telescope (50-80mm f/5).
2. Good equatorial mount with accurate tracking. Guide port available would be nice if you want to start guiding.
3. DSLR or dedicated astro-camera.

You need tools like an adjustable wrench, some screwdrivers, and a level (if you have an equatorial stand). A star chart and a red flashlight are good to have.

Most importantly, you need dark and steady skies.

You can buy used equipment. In fact, I usually do. I purchased my 90mm refractor used and I even purchased my DSLR camera heavily used. Guess what? They both work great. Buying used is a great way to save money.

Part 5: All About You

Please fill the following tables with information about your specific telescopes. These tables will keep you from doing these calculations over and over again.

Telescope 1:

Focal Length of OBJ:

Eyepiece	Magnification	AFOV (deg)	TFOV (deg)

Telescope 2:

Focal Length of OBJ:

Eyepiece	Magnification	AFOV (deg)	TFOV (deg)

Telescope 3:

Focal Length of OBJ:

Eyepiece	Magnification	AFOV (deg)	TFOV (deg)

Telescope 4:

Focal Length of OBJ:

Eyepiece	Magnification	AFOV (deg)	TFOV (deg)

Telescope 5:

Focal Length of OBJ:

Eyepiece	Magnification	AFOV (deg)	TFOV (deg)

Please fill out the following tables with the appropriate information. This enables you to simply write down what you did to be able to successfully see an object and recall that information later when you try again. I have already added the objects you should try to find first with a small telescope.

Object	Magnification	Filter
Jupiter		
Saturn		
Venus		
Mars		
Mizar + Alcor		
Epsilon Lyrae		
Albireo		
M42		
M41		
M13		
Moon		

Object	Magnification	Filter

Object	Magnification	Filter

Object	Magnification	Filter

Object	Magnification	Filter

Object	Magnification	Filter

Take some notes on what you did to obtain successful astrophotos.

Object	Instrument	Settings
		ISO: Shutter: Frames:
		ISO: Shutter: Frames:
		ISO: Shutter: Frames:
		ISO: Shutter: Frames:
		ISO: Shutter: Frames:
		ISO: Shutter: Frames:
		ISO: Shutter: Frames:
		ISO: Shutter: Frames:
		ISO: Shutter: Frames:
		ISO: Shutter: Frames:
		ISO: Shutter: Frames:

Object	Instrument	Settings
		ISO: Shutter: Frames:
		ISO: Shutter: Frames:
		ISO: Shutter: Frames:
		ISO: Shutter: Frames:
		ISO: Shutter: Frames:
		ISO: Shutter: Frames:
		ISO: Shutter: Frames:
		ISO: Shutter: Frames:
		ISO: Shutter: Frames:
		ISO: Shutter: Frames:
		ISO: Shutter: Frames:

Object	Instrument	Settings
		ISO: Shutter: Frames:
		ISO: Shutter: Frames:
		ISO: Shutter: Frames:
		ISO: Shutter: Frames:
		ISO: Shutter: Frames:
		ISO: Shutter: Frames:
		ISO: Shutter: Frames:
		ISO: Shutter: Frames:
		ISO: Shutter: Frames:
		ISO: Shutter: Frames:
		ISO: Shutter: Frames:

Object	Instrument	Settings
		ISO: Shutter: Frames:
		ISO: Shutter: Frames:
		ISO: Shutter: Frames:
		ISO: Shutter: Frames:
		ISO: Shutter: Frames:
		ISO: Shutter: Frames:
		ISO: Shutter: Frames:
		ISO: Shutter: Frames:
		ISO: Shutter: Frames:
		ISO: Shutter: Frames:
		ISO: Shutter: Frames:

Object	Instrument	Settings
		Video Length: % Stacked:
		Video Length: % Stacked:
		Video Length: % Stacked:
		Video Length: % Stacked:
		Video Length: % Stacked:
		Video Length: % Stacked:
		Video Length: % Stacked:
		Video Length: % Stacked:
		Video Length: % Stacked:

Object	Instrument	Settings
		Video Length: % Stacked:
		Video Length: % Stacked:
		Video Length: % Stacked:
		Video Length: % Stacked:
		Video Length: % Stacked:
		Video Length: % Stacked:
		Video Length: % Stacked:
		Video Length: % Stacked:
		Video Length: % Stacked:

Object	Instrument	Settings
		Video Length: % Stacked:
		Video Length: % Stacked:
		Video Length: % Stacked:
		Video Length: % Stacked:
		Video Length: % Stacked:
		Video Length: % Stacked:
		Video Length: % Stacked:
		Video Length: % Stacked:
		Video Length: % Stacked:

Object	Instrument	Settings
		Video Length: % Stacked:
		Video Length: % Stacked:
		Video Length: % Stacked:
		Video Length: % Stacked:
		Video Length: % Stacked:
		Video Length: % Stacked:
		Video Length: % Stacked:
		Video Length: % Stacked:
		Video Length: % Stacked:

Additional Notes:

Additional Notes:

Additional Notes:

Additional Notes:

Thank you for purchasing this book. I hope you enjoyed reading it as much as I enjoyed writing it. Enjoy your new hobby.

If you would like to see more pictures I have taken employing the astrophotography techniques discussed throughout this book, please follow me on social media.

@mmdastro